AF698671

Best Practice Guide on Metals Removal from Drinking Water by Treatment

Metals and Related Substances in Drinking Water Series

Best Practice Guide on Metals Removal from Drinking Water by Treatment

Edited by

Prof. Dr Mustafa Ersoz and Dr. Lisa Barrott

Publishing
London • New York

Published by **IWA Publishing**
Alliance House
12 Caxton Street
London SW1H 0QS, UK
Telephone: +44 (0)20 7654 5500
Fax: +44 (0)20 7654 5555
Email: publications@iwap.co.uk
Web: www.iwapublishing.com

First published 2012

British Library Cataloguing in Publication Data
A CIP catalogue record for this book is available from the British Library

ISBN 9781843393849 (Paperback)
ISBN 9781780400686 (eBook)

Contents

About this Best Practice Guide

This Guide is one of a series produced by the International Water Association's Specialist Group on Metals and Related Substances in Drinking Water. It is a state-of-the-art compilation of the range of scientific, engineering, regulatory and operational issues concerned with the removal of metals from drinking water by treatment. It will be of interest to water utility practitioners, health agencies and policy makers.

The Specialist Group is supported by members from 26 European countries, Canada and the United States. It is an active research network and has regularly convened international conferences and seminars. It has close working links with the World Health Organization, the European Commission's Joint Research Centre, Health Canada and the US Environmental Protection Agency. The Specialist Group developed out of COST Action 637 (www.cost.esf.org) and the funding received from COST from December 2006 to November 2010 is duly acknowledged. The Specialist Group web-site is now part of the IWA Water Wiki accessible via www.iwahq.org.

The Guide is supported by a two-day training course. Information about training courses and the Specialist Group in general is available from www.iwahq.org

Dr. Colin Hayes
Chairman
IWA Specialist Group on Metals and Related Substances in Drinking Water
(METEAU)

Preface

The EU drinking water directive sets a range of standards for metals and related substances in drinking water, many of which are concerned with health protection. A number of these standards are very stringent and require compliance to be assessed at the point of use. Because of the difficulties associated with monitoring, historic practices in many countries have concentrated on the quality of water within the distribution network. As a result, the magnitude of problems with some metals and related substances in drinking water is not fully appreciated in all European countries, and the extent and nature of corrective actions differ widely. Standards for metals in drinking water are variously set by many other countries and often derive from WHO Guidelines; standards vary slightly according to local circumstances.

This Best Practice Guide on Metals Removal from Drinking Water describes drinking water standards and regulations, and explains the impact of a range of water treatment processes on metal levels in drinking water. Its objectives are to provide a basis for assessing the extent of problems and to identify appropriate water treatment options. It provides a reasoned guide to the selection of key water treatment processes. Each chapter focuses on a specific water treatment process and has been written by experts in that particular process.

This Guide provides practice-based knowledge for water engineers and scientists in large and small water utilities, regulatory agencies, health agencies and local municipalities (from cities through to small rural communities). The Guide should also be able to support University level teaching in degree schemes that relate to water management.

Acknowledgements

The editors thank all authors for their patient and dedicated contributions to the chapters of this Guide. The tireless efforts of this Guide's technical editor, Mr. Ahmet Aygun, Environmental Engineer, from Selcuk University are acknowledged.

Authors

Prof. Dr. Asher Brenner, Unit of Environmental Engineering, Ben-Gurion University of the Negev, Beer-Sheva 84105, Israel

Dr. Eddo, J. Hoekstra' Institute for Health and Consumer Protection, Joint Research Centre, European Commission, Ispra I-21027, Italy

Dr. Zsuzsanna Bufa-Dörr, National Institute of Environmental Health, Water Safety Department, Hungary

Prof. Dr. Matyas Borsanyi, National Institute of Environmental Health, Water Safety Department, Hungary

Dr. Larry Russell, Reed International Ltd. Berkeley, CA 94704, USA

Prof. Dr. Mehmet Emin Aydin, Dept. of Environmental Engineering, Selcuk University, Konya, 42031, Turkey

Prof. Dr. Zdravka Lazarova, AIT Austrian Institute of Technology GmbH, Dept. Health & Environm., 3430 Tulln, Austria

Dr. Ali Tor, Dept. of Environmental Engineering, Konya University, Konya, Turkey

Dr. Senar Ozcan, Dept. of Environmental Engineering, Konya University, Konya, Turkey

Dr. Ona Gyliene, Institute of Chemistry of the Centre for Physical Sciences and Technology, A. Goštauto 9, 2600 Vilnius, Lithuania

Dr. Magdelena Zabochnicka-Swiatek, Czestochowa University of Technology, Institute of Environmental Engineering, Brzeznicka 60A, 42-200 Czestochowa, Poland

Prof. Dr. Peter E. Holm, Department of Basic Sciences and Environment, University of Copenhagen, Denmark

Dr. Karin Cederkvist, Department of Basic Sciences and Environment, University of Copenhagen, Denmark

Dr. Sabrina Sorlini, Ricercatrice di Ingegneria Sanitaria Ambientale, DICATA – Facoltà di Ingegneria – Università degli Studi di Brescia via Branze 43 – 25123 Brescia, Italy

Dr. Frausta Prandini, Ricercatrice di Ingegneria Sanitaria Ambientale, DICATA – Facoltà di Ingegneria – Università degli Studi di Brescia via Branze 43 – 25123 Brescia, Italy

Dr. D Staniloae, National Research-Development Institute for Industrial Ecology, ECOIND 90-92 Panduri 050663 Bucharest Romania

Todd Russell, Dept. of Civil and Environmental Engineering, Stanford University, Stanford California 94305, USA
Dr. Brian Croll, Consultant (UK)
Dr. Colin Hayes, Swansea University (UK)
Dr. Tom Hall, Principal Consultant – Water Supply (UK)
Dr. Lisa Barrott, MWH (UK)
Professor Mustafa Ersoz, Selcuk University (TR)

Acronyms

AA	activated alumina
AC	activated carbon
AE	anion exchangers
AIXO-IBR	advanced ion-exchange operations with indefinite brine recycling
Al	aluminum
As	arsenic
BAT	Best Available Technology
BFR	brominated flame retardants
Ca	calcium
CB	carbon block
CCF	coagulation-assisted ceramic filtration
CCPP	calcium carbonate precipiation potential
Cd	cadmium
CE	cation exchangers
CMF	ceramic media filtration
CPC	hexadecylpyridinium chloride
CSF	constructed soil filter
CTAB	hexadecyltrimethyl-ammonium bromide
Cu	copper
DAF	dissolved air flotation
DWD	Drinking Water Directive
DBP	disinfection byproducts
DPF	dual porosity filtration
ED	electrodyalisis
EDCs	endocrine disrupting compounds
EDR	electrodyalisis reversal
EDTA	ethylenediaminetetraacetic acid

EPA	Environmental Protection Agency
EU	European Union
F	Fluoride
Fe	Iron
GAC	granular activated carbon
GFH	granular ferric hydroxide
HS	humic substances
IE	ion exchange
IEMB	ion exchange membrane bioreactor
IMF	immersed membrane filtration
LI	Langelier index
LS	lime softening
MAOA	microsand-assisted oxidation-adsorption
MBR	membrane bioreactor
MCL	maximum contaminant (or concentration) levels
MEUF	Micellar-Enhanced Ultrafiltration
Mg	Magnesium
Mn	manganese
MF	microfiltration
MIEX	Magnetically impregnated resins
Ni	nickel
NO_2^-	nitrite
NO_3^-	nitrate
NOM	natural organic matter
NF	nanoflitration
NP	nanoparticles
OMP	organic micro-pollutants
O&M	operations and maintenance
PA	polyamide
PAC	powdered activated carbon
PAC/CFMC	powdered activated carbon/crossflow microfiltration
PAH	polycyclic aromatic hydrocarbons
Pb	lead
PEI	polyethylenimine
PEUF	polyelectrolyte-enhanced ultrafiltration
PIES	polyvinyl electrolyte
POU	point-of-use devices
POE	point-of-entry devices
PPCPs	pharmaceuticals and personal care products
RO	reverse osmosis
SBA	strong-base anion exchange resin
SDWA	Safe Drinking Water Act
SMCL	secondary maximum contaminant levels
TOC	total organic carbon
TDS	total dissolved solids

UF	ultrafiltration
UV	ultra violet
WHO	World Health Organization
Zn	Zinc

Definitions

Activated carbon (AC)	Activated carbon is a natural material derived from bituminous coal, lignite, wood, coconut shell etc. which is activated by heat, steam and other means. Each activated carbon has different adsorption properties.
Adsorption	The process by which chemicals are held on the surface of activated carbon.
Aeration	The process of adding air to water. Air can be added to water by either passing air through water or passing water through air.
Aerobic	A condition in which free (atmospheric) or dissolved oxygen is present in the water.
Autocatalytic reactions	Autocatalytic reactions are chemical reactions in which at least one of the reactants is also a product.
Best Available Technology (BAT)	The water treatment technology that a regulator recognizes as the most effective for removing a contaminant.
Biosorbents	The removal of heavy metals and other hazardous substances by the passive binding to biosorbents such as non-living micro-organisms (algae, fungi and bacteria) and other biomass (peat, rice hull, wheat shell, fruit peel, leaves, saw dust, bark of trees, macrofungus etc.) from an aqueous solutions.
Brominated flame retardants (BFRs)	Organobrominated compounds which have been routinely been added to consumer products for several decades to reduce fire-related injury and property damage.
Calcium Carbonate Precipitation Potential (CCPP)	The theoretical concentration of calcium carbonate that could precipitate from a water.
Chemical oxidation	A chemical process that causes the loss of electrons from an element or ion. In water treatment chemical oxidation may be achieved by aeration or the addition of a chemical such as chlorine or ozone.

Chemical Precipitation	A common technology used to remove dissolved (ionic) metals from water. The ionic metals are converted to an insoluble form (particle) by the chemical reaction between the soluble metal compounds and the precipitating reagent. The particles formed by this reaction are removed from solution by settling and/or filtration.
Coagulation	The clumping together of very fine particles into larger particles caused by the use of chemicals (coagulants). The chemicals neutralize the electrical charges of the fine particles and cause destabilization of the particles. The coagulant is added during a period of intense agitation.
Electrochemical treatment	An emerging technology used for the removal of organic and inorganic impurities from water and wastewater. They involve redox reactions, where oxidation and reduction reactions are separated in space or time.
Electrocoagulation	The electrochemical production of destabilisation agents (usually Al, Fe ions) that neutralize the electric charge of suspended pollutant.
Electrodialysis (ED) and Electrodialysis Reversal (EDR)	Electrodialysis (ED) is used to remove substances possessing charge from solution through ion-exchange membranes under the influence of the applied electric potential difference. In EDR, the direction of ion flow is reversed periodically by reversing the polarity of the applied electric current.
Electrolysis	A process in which one species in solution (usually a metal ion) is reduced by electrons at the cathode and another gives up electrons to the anode and is oxidized there.
Endocrine disrupting compounds (EDCs)	EDCs include various types of natural and synthetic chemical compounds which mimic or inhibit the reproductive action of the endocrine system in animals and humans.
Enhanced coagulation/filtration	Enhanced coagulation involves modifications to an existing coagulation process such as increasing the coagulant dosage, reducing the pH, or both. The enhanced coagulation process is followed by filtration.
Enhanced lime softening	Lime softening is a process where hydrated lime or quicklime is added to raise pH and precipitate calcium. In enhanced softening, the pH is increased further in a second stage, to at least 10.6 to also remove magnesium and some arsenic.
Flocculation	The gathering together of fine particles in water by gentle mixing after the addition of coagulant chemicals to form larger particles.
Flotation/Dissolved Air Flotation (DAF)	A water treatment process that clarifies wastewaters water by the removal of suspended matter such as oil or solids. The removal is achieved by dissolving air in the water under pressure and then releasing the air at atmospheric pressure in a flotation tank.
Filtration	A process for removing particulate matter from water by passage through porous media.
Ion exchange (IX)	A physical/chemical process by which an ion on the solid phase is exchanged for an ion in the feed water. This solid phase is typically a synthetic resin which has been chosen to preferentially adsorb the particular contaminant of concern.

Langelier Index (LI)	An index reflecting the equilibrium pH of a water with respect to calcium and alkalinity; used in stabilizing water to control both corrosion and scale deposition.
Maximum contaminant (or concentration) levels (MCL)	The highest level of a contaminant that the USEPA allows in drinking water. MCLs ensure that drinking water does not pose either a short-term or long-term health risk. The USEPA sets MCLs at levels that are economically and technologically feasible.
Organic micro-pollutants (OMPs)	Any organic compounds that may be found at microgram per litre concentrations or lower in water, such as pesticides, pharmaceutical residues, hormones, flame-retardants, plasticizers, perfluorinated compounds, and others.
Oxidation/Filtration	Oxidation/filtration refers to precipitative processes that are designed to remove naturally occurring iron and manganese from water. The processes involve the oxidation of the soluble forms of iron and manganese to their insoluble forms and then removal by filtration.
Pharmaceuticals and personal care products (PPCPs)	PPCPs comprise a diverse collection of thousands of chemical substances, including prescription and over-the-counter therapeutic drugs, veterinary drugs, fragrances, and cosmetics. In general PPCPs include products used by individuals for personal health or cosmetic reasons or used by farmers to enhance growth or health of livestock.
Polycyclic aromatic hydrocarbons (PAHs)	PAHs are a group of persistent organic compounds, some of which are toxic and/or possible or proven human carcinogens; they are produced via incomplete combustion of carbon containing fuels from industrial, commercial, vehicular and residential sources. PAHs in tap water will mainly be due to the presence of PAH-containing materials in water storage and distribution systems.
Reverse Osmosis (RO)	RO is the oldest membrane technology, traditionally used for the desalination of brackish water and sea water. RO produces nearly pure water by maintaining a pressure gradient across the membrane greater than the osmotic pressure of the feed water.
Sedimentation	A water treatment process in which solid particles settle out (by gravity) of the water being treated in a large clarifier or sedimentation tank.
Zeolites	Naturally occurring structured minerals with high cation exchange and ion adsorption capacity.

Foreword

Metals in drinking water can arise from a number of sources. They may occur naturally in source waters because of geological conditions, or result from man-made contamination, for example from mining wastes. They may originate from chemicals used in water treatment, particularly the iron and aluminium based coagulants. The most common cause of lead and copper in drinking water is their use in plumbing systems. Some metals, such as arsenic and lead, have obvious toxicity implications for consumers, and need to be removed effectively to protect public health. Other metals, such as copper or zinc, can impart unpleasant taste to the water at elevated levels, and some, such as iron or manganese, can impair the appearance of the water.

Overall, there is a need to maintain effective water treatment to remove metals from water or prevent contamination during or after treatment. A wide range of water treatment processes are available, and process selection, design and operation for metals removal can pose a challenge to water engineers. This Guide reviews the treatment processes that can be used for metals removal. It covers not only the conventional treatments that have been used for many years, such as coagulation, clarification and filtration, but also membrane processes, adsorption and electrochemical treatment. It therefore provides a comprehensive source of information on treatment for both current and future implementation, to help ensure effective removal of metals from drinking water.

Dr. Tom Hall
Principal Consultant – Water Supply
WRc (UK)

Executive Summary

This Best Practice Guide on Metals Removal from Drinking Water by Treatment describes drinking water standards and regulations for metals, and explains the impact of a range of water treatment processes on metal levels in drinking water. Among water pollutants, heavy metals are considered to cause particular concern because of their cumulative toxic characteristics even at low concentrations. Future drinking water quality standards may probably be gradually changed due to the cumulative knowledge gathered regarding the effect of various substances on human health. Another issue that will affect future trends is the advance of analytical instruments capable of identifying compounds in the nano-gram per liter range, and of other sophisticated sensors for toxicity testing.

Different raw water quality characteristics and operating practices, technical, economical, environmental, geological differences between the countries mean that no single technology is most efficient for metal removal. In choosing a treatment technology attention should be paid to the practicality, efficiency, chemical interferences, public health aspects and possible future improvements.

The process of metal removal includes disposal of waste solids (e.g. spent filter material, coagulant sludge) and residual fluids likely to contain high levels of metals and possibly other hazardous or radioactive (e.g. uranium, radon) constituents. If the water requires pretreatment (e.g. lowering of pH for adsorption processes, or raising it for lime-softening) the waste situation becomes more complicated.

The most common metals of concern in drinking water are iron and manganese, which present similar challenges when found in solution, often appear in tandem. Chemical oxidation converts iron and manganese to insoluble hydroxides and oxides, and the precipitates can then be extracted from solution by sedimentation or filtration. The common oxidants used for iron and manganese removal are oxygen, chlorine, chlorine dioxide, ozone and potassium permanganate. The effectiveness of oxidation on iron and manganese hinges on several key factors: the pH of the solution, the reductive potential of the solution, the presence of naturally occurring organic material (NOM), and the ionic strength of the solution. Manganese is substantially more stable than iron in solution and manganese is autocatalytic where iron is not. Due to the complex chemistry involved in the removal of iron and manganese with oxidation, design and testing on the bench and pilot scale are strongly encouraged.

Coagulation and flocculation are used to form precipitates that adsorb natural organic matters and some inorganic materials such as phosphates, arsenic compounds and fluoride. Aluminum sulphate (alum), ferric iron salts, organic polymers (polyelectrolytes) are widely used as coagulants in the water treatment plants. Oxidative chemicals such as chlorine, ozone or potassium permanganate can be added prior to the primary coagulant to improve the coagulation by oxidizing dissolved organics. All the coagulants have an optimum operation pH range. The Guide describes how heavy metals, existing as cations in water, form insoluble hydroxides and carbonates and precipitate. This is achieved usually by adding caustic soda or lime to adjust the pH according to the maximum metal insolubility.

Metal precipitates or metals adsorbed onto coagulated particles can be removed by both sedimentation and flotation. The sedimentation process is based upon the settlement of the suspended solids by gravity and consequently settleable solids are removed from water as sludge at the bottom of a circular or rectangular tank. General parameters used for the design of sedimentation tank are surface loading rate, retention time, tank depth and flow rate and tank. The major advantage of sedimentation is that it uses only gravity to separate the flocs from water and is therefore cost effective with a low energy usage. The major disadvantages of sedimentation are the long separation time required and the large amount of land area required.

The flotation process uses gas bubbles to increase the buoyancy of suspended solids. To obtain efficient flotation, the pH and chemical dosage of the coagulation process should be optimized. The time and degree of agitation used for flocculation process also affect the performance of flotation process.

A typical treatment train consists of an oxidant dose followed by a detention basin and finally rapid filtration or pressure filtration. The differing chemistries and reaction kinetics of iron and manganese has impacts on filter design. For systems utilizing oxidation for removal of iron and manganese, proper selection of the filter media and design of the filtration unit is important. Filter media (type and size), filter area and depth, hydraulic and solids loading rate and backwashing regimes are all important aspects of filter design. Autocatalytic removal of manganese can take place in a filter and can be critical for manganese removal. An investment in filter pilot testing can provide significant.

Electrochemical treatment is an emerging technology used for the removal of organic and inorganic impurities from water and wastewater. Electrochemical processes involve redox reactions, where oxidation and reduction reactions are separated in space or time. Usually the electrochemical treatment of water is concerned with electron transfer at the solution/electrode interface applying an external direct current in order to drive an electrochemical process. Electrocoagulation is the electrochemical production of destabilisation agents (usually Al or, Fe ions) that neutralize the electric charge of suspended pollutant. Electrochemically generated metallic ions from these electrodes can undergo hydrolysis near the anode to produce a series of activated intermediates that are able to destabilize finely dispersed particles present in the water/wastewater to be treated. Electrochemical treatment methods have a future as advanced technologies for additional treatment of potable water domestically and remote areas.

Zeolites are the best natural filter medium available for treatment of water for metals removal. They offer superior performance to sand and carbon filters, giving purer water and higher throughput rates with less maintenance required. The application of zeolites provides an economical means of removing mixed heavy metals from water. Physical and/or chemical regeneration of used zeolitic adsorbents facilitate the protection of natural environment, provided that the used regenerant can be disposed of safely. Activated carbons are non-selective sorbents and presence of the competing ions could reduce the efficiency of metal removal. All activated carbon filters, do not naturally reduce the levels of soluble salts (including nitrates), fluoride, and some other potentially harmful minerals like arsenic (unless specially designed) and cadmium. The overall water quality (turbidity or presence of other contaminants) affects the capacity of activated carbon to adsorb a specific contaminant.

The Guide describes ion exchange as a reversible chemical reaction where an ion from solution is exchanged for a similarly charged ion attached to an immobile solid particle. These solid ion exchange particles are either naturally occurring inorganic adsorbents or synthetically produced organic resins. The synthetic organic resins are the predominant type used today because their characteristics can be tailored to specific applications. After using, ion exchangers can be cleaned by chemical or thermal processes and reused.

Membrane separation processes are considered emerging technologies capable of improving on the performance of traditional processes such as granular filtration, precipitation, disinfection, biological treatment, and to assist in complete elimination of emerging pollutants. The Guide describes how, within the broad family of membrane separation processes, there are "loose" membrane-based technologies (Microfiltration (MF) and Ultrafiltration (UF)), and "tight" membrane separation processes (Nanofiltration (NF) and Reverse Osmosis (RO)) which require higher levels of pressure to enable free water passage with simultaneous rejection of particulate and even dissolved materials. The advance in membrane science and technology and their high and reliable separation efficiency, make this class of processes very suitable for the removal of heavy metals and related substances.

The Guide describes how physical and chemical processes are applied for arsenic removal in drinking waters: chemical co-precipitation, adsorption, ion exchange, and membrane filtration. Reduced inorganic As(III) should be converted to As(V) to improve its removal and a high Fe:As ratio promotes arsenic removal. The most effective method for arsenic removal in the case of large-scale water utilities is the coagulation with ferric or alumina salts. In addition, adsorption and ion exchange are becoming increasingly popular for arsenic removal in small-scale treatment systems.

The integration of different physical, chemical and/or biological processes into the so-called "hybrid" process creates new advantageous applications in the removal of metals and related substances from ground and surface water. The most known hybrid application is the combination of pressure driven membrane processes such as MF and UF with high capacity sorbents or special metal-bonding agents (polyelectrolytes, surfactants, polyamino acids, natural biopolymers such as chitosan) which substantially increases the separation selectivity and efficiency. The membrane bioreactor process is the best example of the success of hybrid membrane treatment.

Chapter 1

Drinking water quality standards and regulations

Asher Brenner and Eddo J. Hoekstra

1.1 INTRODUCTION

Water is the most essential component for human life. Supply of safe drinking-water is a basic human right that not only protects human health, but also contributes to social development and to the reduction of expenses related to health care. The main aim of establishing this guideline is to assist ensuring the safety of drinking water supplies through the elimination or reduction of problematic constituents that may cause adverse health effects. In this regard, heavy metals are considered to cause particular concern because of their cumulative toxic characteristics even at low concentrations. In general, the purpose of water treatment is to protect the consumer from pathogens, toxic materials and other impurities in the water that may be injurious to human health. Treatment processes can be classified into physical, chemical, and biological processes. Physical treatment processes cause change in contaminants levels through the application of physical forces. In chemical treatment the removal of contaminants is caused by the addition of chemicals or by chemical reactions. In biological treatment the removal of contaminants is caused by biological means such as bacterial degradation. Various treatment processes can be applied, solely or in series, to accomplish gradual and complete removal (in a logical economical manner) of all impurities that pose potential health threats. Water parameters (not all of which are considered contaminants) can be divided into four classes: Physical parameters (such as turbidity, color, taste and odor), sometimes refer to as indicator parameters or organoleptic parameters; Chemical parameters (such as most ions including heavy metals, pesticides and other organic substances); Biological parameters (such as bacteria, viruses, protozoa, and others), and Radiological parameters (including natural and man-made radionuclides). Detailed quantitative standards or recommendations for the various classes of water parameters are given either, by national legislation or guidelines, or by global organizations such as the World Health Organization (WHO, 2004; 2006; 2008).

1.2 DRINKING WATER QUALITY STANDARDS

The purpose of water treatment is to ensure reliable long-term achievement of water quality having concentrations lower than the standards set to avoid any health risk to consumers. Basic water quality requirements include three classes as presented in Table 1.1: aesthetic requirements, human health related requirements, and economical requirements. These basic requirements are translated into specific quantitative standards based on direct studies on human populations, especially related to microbial contamination, and indirect extrapolation obtained from laboratory experiments with target animals. Most numerical values set to be standards are based on risk management considerations that can vary in various locations. Risk considerations refer to health tolerance for a specific contaminant consumed over a lifetime. Contaminants required to be regulated in most countries are divided into several classes, including: microbial substances (such as bacteria, viruses, and parasites); inorganic substances (including heavy metals); organic substances (including pesticides, polycyclic aromatic hydrocarbons (PAH), volatile organic chemicals, and disinfection by-products), and radionuclides. Additional definitions such as water appearance (usually related to turbidity and color) and indicator parameters (which are not directly related to health problems) are also common in various guidelines. Another definition used (for instance by the US EPA) is Secondary Standards, which are non-enforceable guidelines regulating contaminants that may cause cosmetic effects (such as skin or tooth discoloration) or aesthetic effects (such as taste, odor, or color) in drinking water. Most drinking water regulations define a maximum contaminant (or concentration) levels (MCL) for the various substances to be delivered in the water supplied to consumers. Several national regulations also set or recommend treatment technologies (sometimes defined as best available technology – BAT) to achieve the required MCLs for specific contaminants.

Table 1.1 Basic requirements for drinking water quality.

Aesthetic requirements	✓ Turbidity ✓ Color ✓ Taste ✓ Odor
Human health related requirements	✓ Elimination of pathogens and toxic substances ✓ Limitation of problematic ions ✓ Limitation of radionuclides
Economical requirements	✓ Hardness ✓ Corrosion ✓ Iron and manganese

It is important to note, that turbidity which has been always considered as an aesthetic parameter only, is considered today a significant surrogate indicator of water contamination by microorganisms such as *Giardia* and *Cryptosporidium* (Morris *et al.* 1996). Turbidity is therefore commonly serving as a control parameter in many treatment technologies such as granular and membrane filtration, disinfection, desalination, and stabilization. Corrosion on the other hand, which is considered an economical requirement that has major effects on degradation and cost of metal pipes and structures, is considered also as a significant health related problem because of its effect on the release of toxic metals into drinking water systems. While there is a strict requirement to eliminate completely presence in drinking water of pathogens and toxic substances such as heavy metals or organic micro-pollutants (including pesticides, solvents, disinfection by-products, and others), most natural ions are considered essential to human health and their presence in drinking water fulfill partly the recommended dietary daily intake. However, several problematic ions such as nitrate (NO_3^-), nitrite (NO_2^-), and fluoride (F^-), have a maximum concentration limit (MCL) to avoid health effects. On the other hand, there is a minimum concentration limit required in several countries and US states for essential ions that support human health. Fluoride (F^-) is the most known substance still added in many water supply systems. However, other substances such as calcium (Ca^{2+}) and magnesium (Mg^{2+}), have been proposed to be considered as essential additives in water supplies having low mineral content such as desalinated water (WHO, 2009). It should be noted in this regard, that in 1973 the High Council of The Netherlands decided that the Dutch authorities had no legal basis adding chemicals to drinking water if they will not improve the safety as such. This implicates that addition of "medicines" such as fluoride are not allowed. The simple reason is that consumers cannot chose for a different tap water.

1.3 DRINKING WATER LEGISLATION

The objective of the EU Drinking Water Directive (DWD) is to protect the health of the consumers in the European Union and to make sure the water is wholesome and clean. This applies to all water intended for human consumption, as well as to water used in the production and marketing of food. The first revision of the DWD (EU, 1998) sets quality standards for drinking water quality at the tap and obliges Member States to regular monitoring of drinking water quality and to provide to consumers adequate and up-to-date information on their drinking water quality. To make sure drinking water everywhere in the EU is healthy, clean and tasty, the DWD sets standards for the most common parameters that can be found in drinking water. In the DWD a restricted number of parameters which include only essential and health related constituents must be monitored and tested regularly. The WHO guidelines for drinking water are used as a basis for the standards in the DWD. While translating the DWD into national legislation, the Member States of the EU can include additional requirements regarding substances that are relevant within their territory or set higher standards. But Member States are not allowed to set lower standards since the level of protection of human health should be the same within the whole EU. In 2007, the five-yearly revision process as foreseen by the Directive, started and may lead to a second revision (EU, 1998).

The Safe Drinking Water Act (SDWA) is the main federal law that ensures the quality of Americans' drinking water (US EPA, 1996). Under SDWA, the Environmental Protection Agency (EPA) sets standards for drinking water quality and oversees the states, localities, and water suppliers who implement those standards. The SDWA was originally passed by the American Congress in 1974 to protect public health by regulating the nation's public drinking water supply. The law was amended in 1986 and 1996 and requires many actions to protect drinking water and its sources: rivers, lakes, reservoirs, springs, and ground water wells. SDWA authorizes the EPA to set national health-based standards for drinking water to protect against both naturally-occurring and man-made contaminants that may be found in drinking water. Originally, SDWA focused primarily on treatment as the means of providing safe drinking water at the tap. The 1996 amendments greatly enhanced the existing law by recognizing source water protection, operator training, funding for water system improvements, and public information as important components of safe drinking water. This approach ensures the quality of drinking water by protecting it from source to tap. SDWA applies to every public water system in the United States.

1.4 BACTERIOLOGICAL AND MICROBIAL STANDARDS

Waterborne and water related diseases have been always considered among the most serious health problems in the world. Microbial problems in drinking water originate from three sources: feces, environment, and growth in distribution network. Standards for bacteriological and microbial water quality are important means to prevent transmission of disease by pathogenic organisms in water. Most national standards generally follow the recommendations of the WHO. The control of all possible specific pathogens is complex, costly and time consuming. Instead index and indicator parameters have been developed that focus on the detection of fecal pollution, which is the most prominent microbiological problem. Total coliform bacteria is an appropriate indicator to check the effectiveness of water treatment. Escherichia coli is a good general index for the presence of fecal bacteria. Intestinal enterococci as fecal index parameter have an added value to E. coli, because they are more resistant to chlorine disinfection. For cyanobacteria, viruses (e.g. entervirusses) and protozoa (e.g. Cryptosporidium and Giardia) such index and indicator system could not be developed yet. The quality of water leaving the treatment plants and entering the distribution system may deteriorate within the distribution system. This can happen due to access from air valves, defective service reservoirs, cross connections, or un-careful plumbing repairs. A specific problem can occur in the hot water domestic distribution system when the water is not heated above 60°C and when the distribution system has dead volumes. Legionella species can grow under such conditions.

1.5 CHEMICAL, PHYSICAL AND RADIOLOGICAL STANDARDS

Chemical and physical characteristics of water are important because each one affects water use in some manner. Some of them (including radiological properties) are most important due to their direct effect on human health. Some physical characteristics of water such as temperature and pH affect the rate and efficiency of reactions and treatment processes. Presence of suspended solids and colloidal substances deteriorate disinfection processes. The equilibrium of the carbonate system in water has a significant effect on metal pipes and surfaces and may lead to two opposite phenomena (scaling on the one hand and corrosion on the other). There are some natural constituents in water such as arsenic and uranium that have direct proven effect on human health. Other problematic constituents such as heavy metals, pesticides, and radio-nuclides may enter water systems due to anthropogenic activities. The standards determined by global and national authorities set individual levels for each constituent in these classes of materials, to avoid human health problems. Specific standard values for heavy metals and related substances applied by the WHO, EPA and EU, are given in Table 1.2.

Table 1.2 Standards for heavy metals and related substances in drinking water (in mg/L).

Substance	Symbol	Potential health effects from ingestion of water*	WHO	EPA	EU
Aluminium	Al	/	0.1–0.2**	0.05–0.2**	0.2
Antimony	Sb	Increase in blood cholesterol; decrease in blood sugar	0.02	0.006	0.005
Arsenic	As	Skin damage or problems with circulatory systems; increased risk of cancer	0.01	0.01	0.01
Barium	Ba	Increase in blood pressure	0.7	2	/
Boron	B		0.5	/	1
Cadmium	Cd	Kidney damage	0.003	0.005	0.005

(*Continued*)

Table 1.2 Standards for heavy metals and related substances in drinking water (in mg/L) (*Continued*).

Substance	Symbol	Potential health effects from ingestion of water*	WHO	EPA	EU
Chromium	Cr	Allergic dermatitis	0.05	0.1	0.05
Copper	Cu	Short term exposure: Gastrointestinal distress; Long term exposure: Liver or kidney damage	2	1.3	2
Iron	Fe	/	/	0.3	0.2
Lead	Pb	Infants and children: Delays in physical or mental development; children could show slight deficits in attention span and learning abilities; Adults: Kidney problems; high blood pressure	0.01	0.015	0.01
Manganese	Mn	/	0.4	0.05	0.05
Mercury	Hg	Kidney damage	0.001	0.002	0.001
Molybdenum	Mo	/	0.07	/	/
Nickel	Ni	/	0.02	/	0.02
Uranium	U	Increased risk of cancer, kidney toxicity	0.015	0.03	/

*Adapted from the EPA SDWA.
**for large and small installations, accordingly.

1.6 TRENDS FOR THE FUTURE

Future drinking water quality standards may probably be gradually changed due to the cumulative knowledge gathered regarding the effect of various substances on human health. Another issue that will affect future trends is the advance of analytical instruments capable of identifying compounds in the nano-gram per liter range, and of other sophisticated sensors for toxicity testing.

The comprehensive reuse of treated wastewater in many parts of the world or their disposal into receiving streams, may ultimately cause a long-term buildup of toxic chemicals in the closed cycle of water supply and wastewater reuse (Kolpin *et al.* 2002; Ashton *et al.* 2004; Heberer & Adam, 2004; Kim *et al.* 2007). In addition to the "traditional" toxic micro-pollutants such as heavy metals, polycyclic aromatic hydrocarbons (PAHs), and pesticides, there is a growing problem of emerging organic micro-pollutants (OMPs), including pharmaceuticals and personal care products (PPCPs), some of which are considered endocrine disrupting compounds (EDCs). The occurrence of EDCs in the environment showing estrogenic-endocrine modulating effects in aquatic organisms is a "hot" issue of major health-related concern worldwide (Pikering & Sumpter, 2003). Although the extent of threat to humans, being exposed to these EDCs, still remains to be elucidated, their limited biodegradation and high persistence, turns them into a major issue of concern (Heberer & Adam, 2004).

Another emerging group of substances that may pose human health risks is the flame retardants, especially brominated flame retardants (BFRs). Many BFRs have been found to be toxic (acute and chronic), persistent, and of bioaccumulative nature in the environment (Birnbaum & Staskal, 2004). Despite their wide distribution in water, air, food, and human blood, only limited information is available concerning the effect of BFRs on wildlife and man, their environmental fate, and biodegradability potential (Segev *et al.* 2009).

Additional group of compounds including charged inorganic compounds such as perchlorate and bromate, have been found in potentially health risk concentrations in many water sources. Their removal from water especially in the presence of nitrates and other impurities, represent a technological challenge (Velizarov *et al.* 2008).

An opposite future trend is related to the increasing use of desalinated water for domestic water supplies, in many countries facing water scarcity. In desalinated water, the levels of alkalinity and essential minerals, such as calcium and magnesium, are very low (Lahav & Birnhack, 2007). Therefore, desalinated water may be associated with inferior taste, and corrosion problems that result in the release of metal colloids (including heavy metals) into water distribution pipes. In addition, the water-intake of these essential nutrients might be reduced dramatically in some populations. Water treatment processes can affect mineral concentrations and contribute to the total intake of calcium and magnesium for some individuals. Water stabilization based on dissolution of calcium carbonate can supply the required substances to prevent corrosion, including sufficient levels of alkalinity and calcium, and buffering capacity measured by Langelier Index (LI) or by the Calcium Carbonate Precipitation Potential (CCPP). This technology actually covers the need to account for nutritional supply of calcium. However, as evidenced recently (WHO, 2009), a requirement for magnesium addition in desalinated water may be also considered. Other requirements may rise with time.

A specific challenge is to improve the quality of metallic materials releasing less metal ions into drinking water. This is specifically valid for materials that are used in the domestic distribution system and where water has longer stagnation times. Typical examples are release of nickel from chromium-plated taps and removal of the zinc layer of galvanised steel. At EU level, progress is made to harmonise and standardise test methods for materials in contact with drinking water. A next step is to arrive at common acceptance criteria.

A general challenge in drinking water regulation is to implement the WHO-concept of water safety plans, an approach of risk assessment and risk management. The water safety plans cover three basic aspects. Assessment of the system aims to check if the water supply system can deliver safe drinking water. The major part of the assessment is to identify the potential hazards from source to tap. The second basic part is the identification of measures that can control the identified risks. These measures will be mainly the set-up and maintenance of an operational monitoring programme of the water supply and establishing the quality criteria. The third part is the management plans that document the water supply, procedures during normal operation and incident conditions, supporting programmes and communication.

KEY POINTS

The purpose of water treatment is to ensure reliable long-term achievement of water quality having constituent concentrations lower than the standards set to avoid any health risk to consumers. Water parameters (not all of which are considered contaminants) can be divided into Physical, Chemical, Biological, and Radiological parameters. Detailed quantitative standards or recommendations for the problematic water parameters are given either, by national guidelines, or by global organizations such as the World Health Organization (WHO). Most drinking water regulations define a maximum contaminant (or concentration) levels (MCL) for the various substances to be delivered in the water supplied to consumers. Several regulations also set or recommend treatment technologies (sometimes defined as best available technology – BAT) to achieve the required MCLs for specific contaminants. Among water pollutants, heavy metals are considered to cause particular concern because of their cumulative toxic characteristics even at low concentrations.

REFERENCES

Ashton D., Hilton M. and Thomas K. V. (2004). Investigating the environmental transport of human pharmaceuticals to streams in the United Kingdom. *Sci. Total Environ.*, **333**, 167–184.

Birnbaum L. S. and Staskal D. F. (2004). Brominated flame retardants: cause for concern? *Environ. Health Perspect.*, **112**, 9–17.

EU (1998). Council Directive 98/83/EC of 3 November 1998 on the quality of water intended for human consumption. *Official Journal of the European Union*, **L330**, 32–54. http://ec.europa.eu/environment/water/water-drink/index_en.html.

Heberer T. and Adam M. (2004). Transport and attenuation of pharmaceutical residues during artificial groundwater replenishment. *Environ. Chem.*, **1**, 22–25.

Kim S. D., Cho J., Kim I. S., Vanderford B. J. and Snyder S. A. (2007). Occurrence and removal of pharmaceuticals and endocrine disruptors in South Korean surface, drinking, and waste waters. *Water Res.*, **41**(5), 1013–1021.

Kolpin D. W., Furlong E. T., Meyer M. T., Thurman E. M., Zaugg S. D., Barber L. B. and Buxton H. T. (2002). Pharmaceuticals, hormones, and other organic wastewater contaminants in U.S. Streams, 1999–2000: a national reconnaissance. *Environ. Sci. Technol.*, **36**(6), 1202–1211.

Lahav O. and Birnhack L. (2007). Quality criteria for desalinated water following post treatment. *Desalination*, **206**, 286–303.

Morris R. D., Naumova E. N., Levin R. and Munasinghe R. L. (1996). Temporal variation in drinking water turbidity and disguised gastroenteritis in Milwaukee. *Am. J. Public Health*, **86**(2), 237–239.

Pikering A. S. and Sumpter J. P. (2003). Comprehending endocrine disrupters in aquatic environments. *Environ. Sci. Technol.*, **37**(17), 331–336.

Segev O., Kushmaro A. and Brenner A. (2009). Environmental impact of flame retardants (persistence and biodegradability). *Int. J. Environ. Res. Public Health*, **6**(2), 478–491.

US EPA (1996). Safe Drinking Water Act (SDWA). http://www.epa.gov/safewater/sdwa/index.html.

Velizarov S., Matos C., Oehmen A., Serra S., Reis M. and Crespo J. (2008). Removal of inorganic charged micropollutants from drinking water supplies by hybrid ion exchange membrane processes. *Desalination*, **223** (1–3), 85–90.

WHO (2004). Guidelines for Drinking-Water Quality. 3rd edn, World Health Organization, Geneva, p. 515.

WHO (2006). Guidelines for Drinking-Water Quality. First Addendum to 3rd edn, World Health Organization, Geneva, p. 595.

WHO (2008). Guidelines for Drinking-Water Quality. Second Addendum to 3rd edn, World Health Organization, Geneva, p. 103.

WHO (2009). Calcium and Magnesium in Drinking-water: Public Health Significance. World Health Organization, Geneva, p. 180.

Chapter 2

Guide to the selection of water treatment processes for removal of metals

Zsuzsanna Bufa-Dőrr, Mátyás Borsányi and Ali Tor

2.1 INTRODUCTION

The chemistry, the potential health risk of the different metal substances and their natural occurrence in some water resources sometimes make it necessary to remove metals from drinking water. The different raw water quality characteristics, operating practices and the technical, economical, environmental and geological differences between countries mean that there is no single technology that is most efficient for all applications. In choosing a treatment technology, attention should be paid to the practicality, efficiency, chemical interferences, public health aspects and possible future improvements.

2.2 TECHNOLOGIES FOR REMOVE METALS FROM DRINKING WATER

The most common heavy metals that have been identified in polluted water are arsenic, copper, cadmium, lead, chromium, nickel mercury and zinc. The release of these metals with no suitable treatment threatens public health due to their persistence in the environment and accumulation in the food chain (Akpor & Muchie, 2010). According to Hashim *et al.* (2011), lead exists in the environment in 0 and +2 oxidation states. Pb(II) is the more common and reactive state of Pb. The complexation of Pb with organic (humic and fulvic acids, EDTA, amino acids) and inorganic ligands, including Cl^-, $CO_3{}^{2-}$, $SO_4{}^{2-}$, $PO_4{}^{3-}$ results in its low solubility compounds. Chromium occurs in 0, +6 and +3 oxidation states. Hexavalent form [Cr(VI)] is the dominant and toxic form of Cr in waters. At low pH, especially at $pH < 4$, Cr(III) becomes the dominant form of Cr in the aqueous phase. Zinc presents in 0 and +2 oxidation states in environment. At high pH, Zn is bioavailable. Cadmium can occur in 0 and +2 oxidation states. It precipitates in the presence of phosphate, arsenate, chromate, sulphide, and so on. It has mobility at pH range 4.5–5.5. Cupper can occur in 0, +1 and +2 oxidation states. The most toxic species of Cu is cupric ion (Cu^{2+}), including $Cu(OH)^+$ and $Cu_2(OH)_2{}^{2+}$. In aerobic alkaline aqueous phase, the dominant soluble species is $CuCO_3$. Arsenic can occur in both organic and inorganic forms in natural waters. The oxidation-reduction conditions and the pH of the water affect the valence and species of inorganic arsenic. The reduced, trivalent form [As(III)] normally is found in ground water (anaerobic conditions) and the oxidized, pentavalent form [As(V)] is found in surface water (aerobic conditions), but sometimes in the ground water both forms have been found together in the same water source.

Although soluble arsenic species typically make up the majority of the total arsenic concentration in natural waters, some research indicates that arsenic can exist as particulate at significant concentrations. (Fields *et al.* 2000).

Many techniques that have been used to remove heavy metal ions from drinking water include chemical precipitation, ion-exchange, adsorption and membrane filtration, and so on. This chapter deals with these available technologies and their advantages as well as limitations in applications. Due to the oxidation states of arsenic, selection of an arsenic removal technique is more complex than those used for other heavy metal ions.

A summary of the possible techniques for metal removals, with their advantages and disadvantages, is given in Table 2.1. These processes are described in more detail in the following chapters.

Table 2.1 Comparative overview of the treatment technogies used for removing heavy metals and arsenic from water (Khandaker 2009; Hashim *et al.* 2011).

Treatment technology	Institutional experience	Advantage and disadvantage
Coagulation with iron salts and alum	Well proven at centralized plants, piloted at community and household levels.	Relatively inexpensive. Phosphate and silicate may reduce arsenic removal rates. Generates heavy metal-rich sludge.
Iron based surfaces	Well proven at centralized plants for the removal of As	Easily controlled, Relatively inexpensive, not normally regenerated, used material may be classed as toxic waste.
Lime softening	Proven effective in laboratories and at pilot scale. Efficiency of this chemical process should be largely independent of scale. Chiefly seen in central systems in conjunction with water softening.	Extreme pH and large volume of waste generated. More expensive than coagulation with iron salts or alum because of larger doses required, and waste handling.
Ion exchange	Pilot scale in central and household systems, mostly in industrialized countries.	Selectively remove low level of metal ions from contaminated aquifer, despite high concentration of natural component Long-term performance of regenerated media needs documentation. Waters rich in iron and manganese may require pre-treatment to prevent media clogging. Moderately expensive. Regeneration produces metal-rich brine.
Adsorption by activated carbon (AC)	Pilot scale in community and household systems, in industrialized and developing countries.	High BET surface area and surface active agents provide adsorption. Regeneration of spent materials may be frequently needed. Expensive.
Adsorption by industrial by products and wastes	Pilot scale in community and household systems, in industrialized and developing countries.	These are readily available from industry; show promising result, field application needed.
Membrane technology	Shown effective in laboratory studies in industrialized countries.	High removal efficiency observed. Pretreatment usually required. Relatively expensive, especially if operated at high pressures.

2.2.1 Coagulation/Filtration technology

Coagulation processes use either Fe(III) or Al(III) salts to form flocs. When mixed with influent water, iron or aluminum salt precipitating agents form ferric or aluminum hydroxides (Öllős, 1998). Arsenic may be converted to an insoluble form by precipitation, co-precipitation onto the hydroxide phase, or adsorption onto a solid oxyhydroxide surface site (Remembrance & Möller 2006). A big advantage of the method, that it can be combined with removal of iron and manganese and other metals from water. The precipitated $Fe(OH)_3$ or $Al(OH)_3$ can be removed either by granular media filtration or membrane microfiltration. A disadvantage of using granular media filtration is that a flocculation step must be included to facilitate growth of floc particles. In contrast, use of a membrane microfilter can eliminate the need for this step (Khandaker *et al.* 2009). The efficiency of the method depends on pH, dose and type of the coagulant (Fe or Al salt), raw water quality and concentration of metals etc (Öllős, 1998). Addition of a cationic polymer coagulant can also improve the efficiency of flocculation (Remembrance & Möller 2006).

Chemical precipitation is also used in water treatment as an effecient process for metal removal like coagulation and lime softening. This process occurs through the transformation of the heavy metals existing as cations in water into insoluble hydroxides and carbonates by adding chemical precipitants (i.e., lime, NaOH, etc.). This facilitates the contaminant's subsequent removal from the liquid phase by physical methods, that is, filtration. During the treatment, the required amount of chemical precipitant depends on the pH and alkalinity of the water. Efficiency of chemical precipitation depends on the solubility of the complexes formed (Akpor & Muchie, 2010). In hydroxide precipitation process, the addition of coagulants such as alum, iron salts, and organic polymers can enhance the removal of heavy metals from polluted water (Fu & Wang, 2011).

In an arsenic removal technology the optimal pH is less than 7.5. (Khandaker *et al.* 2009). Fe salts are more effective than Al salts in the pH range 7.0–7.5, but at lower pH the removal efficiency is quite similar (Table 2.2) (Öllős, 1998).

Table 2.2 Removal of heavy metals via using different coagulants and lime (Remembrance & Möller, 2006; Pontius, 1990).

Heavy metals	Coagulants	Removal efficiency, %
As(V)	$FeCl_3$	81–100
	$Fe_2(SO_4)_3$	80–99
	$Fe(OH)_3$	94–96
	$Al_2(SO_4)_3$	85–98
	Polyaluminum-chloride	90
	Na_2S	80
	Lime (pH 11)	>90
Cr(III)	$Al_2(SO_4)_3$ (pH 7–9)	>90
	Lime (pH 6.5–9)	>95
Pb	$Fe_2(SO_4)_3$ (pH 6–9)	>95
	$Al_2(SO_4)_3$ (pH 6–9)	>95
	Lime (pH 7–8.5)	>95

Adding lime to soften waters (remove Ca & Mg) often removes appreciable amounts of inorganic pollutants such as As, Cd, Ni, Cu, Pb, Zn through sorptive uptake by metal carbonates and hydroxides (Khandaker *et al.* 2009). Lime softening is most effective when pH is high (greater than 10.5) and chlorine is used to oxidize arsenite to arsenate. As(V) removal is decreased when orthophosphate and carbonate are present (Remembrance & Möller 2006). Adding magnesium-hydroxide is more effective to remove arsenic, than adding sodium carbonate. Removal rates may be increased significantly by adding some iron to the influent. It may be difficult to reduce consistently to 1 μg/L by lime softening alone. Systems using lime softening may require secondary treatment to meet that goal. Table 2.2 also presents the removal efficiency of different coagulants towards heavy metals in water.

The presence of phosphate in the raw water also impacts the effectiveness because the phosphate forms insoluble precipitates with the added coagulant, which reduces the amount of available coagulant for arsenic removal. Furthermore the phosphate and the carbonate/bicarbonate (alkalinity) compete with the arsenate for the free adsorption surface of the hydroxide flakes (Laky & Licskó, 2011; Laky, 2009).

The presence of silica in the raw water inhibits the growth of the flakes and, for example – Fe(III) and the silica form soluble polymers (Laky & Licskó, 2011).

2.2.2 Adsorption technology

Adsorption is seen as effective and economic technique for removal of heavy metal from water. The adsorption process provides flexible operation and in most cases it will give high-quality treatment. Moreover, adsorbents can be regenerated by suitable desorption process if a reversible adsorption occurs (Fu & Wang, 2011).

Although activated carbon is the most widely used adsorbent for water treatment, it is very expensive and has high operating costs due to the high price of the activated carbon and to the high water flow rate always involved, and these costs can be greatly increased when there are no carbon regeneration units locally (Gong *et al.* 2005). Therefore, in recent years, considerable attention has been devoted to the study of different types of low-cost materials in order to remove the heavy metals from polluted waters.

Activated carbon (AC)

Activated carbon (AC) is recognized as chemically stable materials and presents a high adsorption capacity for many heavy metals due to its large surface area and presence of different types of surface functional groups, for example hydroxyl, carbonyl, lactone, carboxylic acid (Fu & Wang, 2001).

Han *et al.* (2000) reported that granular AC is highly suitable for use in permeable barriers, especially for removing Cr(VI) from contaminated groundwater. Regeneration of carbon by phosphate extraction and acid washing is also successful (Han *et al.* 2000), allowing the possibility for repeated use of the granular AC for removing Cr(VI) from water. Commercial activated carbons are extensively used for As (III) and As(V) adsorption from water (Lorenzen *et al.* 1995). A high arsenic sorption capacity (2860 mg/g) was observed on AC (Rajakovic, 1992).

Mesoporous carbon containing iron prepared from a silica template can be used for As removal from drinking water. The maximum adsorption capacities were 5.96 mg/g for arsenite and 5.15 mg/g for arsenate (Dwivedi *et al.* 2008). They also found in column sorption systems that the adsorption capacity of granular activated carbon for Pb(II) is 2.0132 mg/g for a 60 mg/L feed concentration of Pb(II) at hydraulic loading rate of 12 $m^3/(m^2 \cdot h)$ and 0.6 m column bed height.

Iron based surfaces

Iron based surfaces are widely known for their affinity for arsenic, however, not all iron minerals have a high capacity. Amorphous $Fe(OH)_3$ is especially effective and is commercially available as GFH granular ferric hydroxide which is synthesised from high quality raw materials. In Europe, the Bayoxid E33 adsorbent is very common. GFH is roughly 50% Fe-OH and 50% water. Because of its high water content, special care is required in the shipping and handling of GFH. Occasionally, "clumping" of GFH media causes a decrease in permeability and headloss, requiring back-washing to remove fines (Khandaker *et al.* 2009).

This technique combines simple treatment equipment, less maintenance and safe operation over a long period. The GFH adsorption process requires no chemical additives of any kind (Driehaus & Dupont, 2005).

GFH has extremely high adsorption capacities, up to 60 g/kg for arsenic. The media is reported to be more efficient in removal of both arsenite and arsenate than activated alumina (AA) on a wider pH range than AA (Kardos, 2006).

The media is efficient in removing arsenic over a wide range of pH, but if the pH higher than 7.5 arsenate removal falls (Kardos, 2006). At low pH, the adsorption density of arsenic(V) is much higher than of arsenic (III), but at slightly alkaline pH, adsorption is nearly equal for both oxidation states of arsenic (Driehaus & Dupont, 2005). Alkaline pH reduces the lifetime of GFH (Kardos, 2006).

GFH can adsorb the other dissolved substances: phosphate antimony, molybdenum, copper, lead, uranium, selenium, vanadium, which can compete with arsenic on the surface of the material. According to toxicity characteristic the spent GFH can be disposed as non-toxic solid waste (Kardos, 2006). However in some countries it is classed as toxic waste. Moreover, there are a lot of papers describing As removal by different ferrous materials. The reader may refer to Mohan & Pittman (2007) for a detailed discussion on ferrous and other adsorbents for removal of arsenic from water.

Activated alumina AA

AA has been the most commonly used adsorbent for arsenic removal, since it is available as a commercially adsorbent material for water treatment. Activated Alumina (AA) is prepared by partially dehydrating $Al(OH)_3$ at high temperatures (Remembrance & Möller 2006).

The maximum adsorptive capacity of AA is 5 to 24 mg/g media at equilibrium arsenic concentrations of 0.05 to 0.2 mg/L (Remembrance & Möller 2006). Activated alumina is most effective at pH 5.5–6.0. For most applications, acid addition is required to optimize the process (Selecky *et al.* 2005). Disadvantages of the operation of AA systems are the hazardous chemicals required to maintain proper pH, such as HCl, and to regenerate media, such as caustic soda.

AA is highly selective towards As(V) and this strong attraction results in regeneration problems, possibly resulting in 5 to 10 percent loss of adsorptive capacity for each run.

At pH 5.5–8.5 the anion order of preference of AA, from most to least preferred is (Remembrance & Möller 2006): $OH^- > H_2AsO_4^-$, $Si(OH)_3O^- > F- > HSeO_3^- > SO_4^{2-} > CrO_4^{2-} >> HCO_3^- > NO_3^- > Br^- > I^-$

Greensand

Greensand, which contains both Mn and Fe-rich glauconite is a naturally-occurring material. Greensand filtration removes arsenic by oxidative sorption where manganese in the greensand oxidizes any arsenite to arsenate which subsequently sorbs to iron phases in the greensand. When the greensand's oxidative capacity is exhausted potassium permanganate is used to regenerate it. Greensand is not an especially effective adsorbent compared to Fe-based media (Khandaker *et al.* 2009).

Other media

Because AC is very expensive, finding other potential low-cost adsorbents that are able to remove arsenic and other heavy metal ions have become an important issue in water treatment. To date, hundreds of studies on the use of low-cost adsorbents have been published. The wood charcoal, banana pith, coal fly ash, spent tea leaf, mushroom, saw dust, rice husk, sand, water hyacinth, other gricultural wastes and industrial byproducts have been studied as adsorbents for the heavy metal removal from water (Fu & Wang, 2011).

2.2.3 Co-removal of arsenic, iron and manganase

The existence of arsenic is often correlated with high Fe(II)-Mn(II) levels, particularly in ground waters and reducing the concentrations of all three elements may be desirable (Remembrance & Möller 2006).

Co-removal of the metals involves oxidizing reduced iron to remove arsenic through sorption/coprecipitation/coagulation (Öllős, 1998). Oxidation of Fe(II) will lead to the formation of $Fe(OH)_3$ precipitate; arsenic will either be co-precipitated or adsorb to the iron hydroxide form and be removed. Iron as Fe(III) can be added to the untreated water if there is not enough iron for this process to be efficient. Oxidation of Mn(II) without Fe(II) becomes effective for arsenic removal only at influent Mn (II) concentrations greater than 3 mg/L (Remembrance & Möller 2006).

Using aeration or chlorination to oxidize the iron usually causes arsenic removal as well. Researches and practical experiments shows that Al and Fe precipitation processes, through Fe removal technology in conventional water treatment plants, can be optimized for removal of arsenic at low cost (Khandaker *et al.* 2009).

2.2.4 Ion exchange

Ion-exchangers may be either synthetic,usually polymer resins, or natural solids. Ion exchangers are able to exchange its cations with the metals in the water. In the ion-exchange processes, synthetic resins are often preferred because of their effectiveness in the removal of the heavy metals from the solution The strongly acidic resins with sulfonic acid groups and weakly acid resins with carboxylic acid groups are the most common cation exchangers. Hydrogen ions in the sulfonic group or carboxylic group of the resin behave as exchangeable ions with metal cations (Fu & Wang, 2011). The sorption of heavy metal ions by ion-exchange resins is influenced by different operating variables such as pH, temperature, initial metal concentration and contact time (Gode & Pehlivan, 2006).

Apart from synthetic resins, natural zeolites, which are naturally occurring silicate minerals, have been widely used for the removal of heavy metal from water because of their low cost and high abundance. Up to now, many researchers demonstrated that zeolites have good cation-exchange capacities toward heavy metal ions under different experimental conditions. For example, in batch experiments, sorption capacities of zeolites for Pb(II) and Zn(II) are 1.361 and 2.237 meq/g, respectively (Athanasiadis & Helmreich, 2005; Berber-Mendoza *et al.* 2006).

In arsenic removal from water by this process, ion exchange resins can remove As(V) better than As(III) in the pH range 6–9 (Öllős, 1998). The strong-base anion exchange (SBA), sulfate-selective (as opposed to nitrate-selective) resins are best for arsenic removal, the cation-selective resins are inappropriate (Remembrance & Möller 2006; Öllős, 1998).

The anion order of preference of SBA resins, from most to least preferred is (Öllős, 1998): $SO_4^{2-} > HAsO_4^{2-}, HPO_4^{2-} > NO_3^-, CO_3^{2-} > Cl^-, H_2AsO_4^-, HCO_3$

In the process of regeneration by NaCl, the chloride is the anion that is displaced from the resin for arsenic during the operational phase. This process may not be appropriate for some systems that already have high chloride present in influent, as resulting chloride concentrations in product water could lead to corrosion of iron from water supply pipes (Remembrance & Möller 2006).

Sulfate, TDS, selenium, fluoride, and nitrate compete with arsenic and can affect the run length. Passage through a series of columns could improve removal and decrease regeneration frequency. Suspended solids and precipitated iron can cause clogging of the ion excahnge bed. Systems containing high levels of these constituents (especially ground water treatment) may require pretreatment.

2.2.5 Membrane processes

Membrane processes using different types of membranes can be operated for heavy metal removal because of their high efficiency, easy operation and space saving. The membrane processes used to remove metals from the water are reverse osmosis (RO), ultrafiltration (UF), nanofiltration (NF) and electrodialysis (ED). Microfilters (MF) and UF exclude water constituents based on their size. The pore sizes of MF and UF are in the range of (0.1–3) and (0.01–0.1) μm, respectively. NF has a smaller pore size (0.001–0.01 μm) than MF or UF and can therefore exclude significant portions of dissolved heavy metal ions, however it is also more susceptible to fouling than MF or UF (Remembrance & Möller 2006).

Reverse osmosis (RO)

In this process, water passes through the membrane while the dissolved and particulate matter is rejected. The process is very effective for removal of ionic species from solution. The resulting concentrated by-product solutions make recovery of metals more feasible. Using appropriate RO systems to remove heavy metals has been investigated (Table 2.3), but these have yet to be widely applied (Fu & Wang, 2011). For the removal of arsenic, when operating pressure is ideal, reverse osmosis is efficient for removal both arsenite (43–84%) and arsenate (97–99%). The range of particles that RO membranes can retain is 0.005–0.5 μm, including ions. The most effective membrane type is polyamide membrane (PA) (Öllős 1998).

Table 2.3 Some examples of application of RO, NF and RO + NF to the removal of some heavy metals from water.

Membrane process	Heavy metal	Operating conditions	Removal efficiency, %
RO	Cu(II) and Ni(II)	Pressure: 5 atm	99.5
RO	As(III) and As(V)	Not available	As(III): 20–55, As(V): 91–99
RO	Ni(II) and Zn(II)	Pressure: 1100 kPa	Ni(II): 99.3, Zn(II): 98.9
NF	Cu(II)	Flat sheet membrane and pressure: 20 bar	96–98
RO + NF	Cu(II)	Pressure: 35 bar	>95

The pH did not affect removal in the range 4–8, although in the case of cellulose-acetate membrane materials pH should be 5–6.5. Potentially competing anions, including sulfate, also did not affect arsenic removal. A disadvantage to arsenic treatment by RO is the low ratio of treated product water to required inflow (Remembrance & Möller 2006). Using RO causes the treated water to be very corrosive and it should be mineralized.

RO removal efficiency can be 60% of the arsenic from unchlorinated water and 90% from chlorinated water Preoxidation of arsenic by chlorination cause the huge difference in the efficiency (Remembrance & Möller 2006).

Nanofiltration (NF)

NF is the intermediate process between UF and RO. NF is a promising technology for the rejection of heavy metal ions. NF process provide ease of operation, reliability and comparatively low energy consumption as well as high efficiency of pollutant removal (Fu & Wang, 2011). Murthy & Chaudhari (2008) applied a thin-film composite polyamide NF membrane to removal of Ni(II) from aqueous solution The maximum Ni(II) rejection was found as 98% and 92% for initial Ni(II) concentration of 5 and 250 mg/L, respectively. Nanofiltration is capable of arsenic removal of over 90%. The recoveries range is between 15 to 20%, but in constant high quality (Surd 2001). The increased water recovery can lead to increased costs for arsenic removal. The high costs of arsenic removal causes that nanofiltration can be efficient process only in small systems ($<1000\ m^3/d$). Using nanofiltration cause that treated water also should be mineralized. Some examples of application of RO, NF and RO + NF to the removal of some heavy metals from water is listed in Table 2.3 (Fu & Wang, 2011).

Electrodialysis (ED)

As(III) after pre-oxidation can be removed from solution under the influence of an electrical potential difference. Removal efficiency can be 80% for arsenic (Öllős 1998). Tor *et al.* (2004) also reported that more than 90% removal of Cr(VI) from aqueous solution containing 5.2 mg/L of Cr(VI) at current density of $1.4\ mA/cm^2$.

2.2.6 Others

Newer technologies that are being tested include sand ballasted coagulation-sedimentation, fluidized-bed in situ oxidation adsorption, coagulation-assisted ceramic media filtration, and immersed media with carrier particles. Macromolecule heavy metal flocculants is a new kind of flocculant. For example, mercaptoacetyl chitosan, by reacting chitosan with mercaptoacetic acid, could not only remove turbidity, but also remove heavy metals in water Chang *et al.* (2009). Flocculants of Konjac-graft-poly (acrylamide)-co-sodium xanthate and polyampholyte chitosan derivatives-*N*-carboxyethylated chitosans were also used to remove heavy metals from water (Fu & Wang, 2011). Sand ballasted coagulation-sedimentation uses sand and polymer additions to coagulation to boost arsenic removal. Fluidized-bed in situ oxidation adsorption involves adsorbing ferrous iron onto a continuously generated sand surface. Oxidation of the iron leads to uptake and removal of arsenic. Coagulation-assisted ceramic media filtration is similar to CMF except that ceramic filters are used in the floc removal step. Immersed media with carrier particles is an in situ filtration process (Khandaker *et al.* 2009).

2.3 POINT OF USE/POINT OF ENTRY (POU/POE) DEVICES

The Point-of-Use (POU) and Point-of-Entry (POE) devices as options can be used for removal of heavy metal ions from drinking water. POU and POE devices can be effective and affordable compliance options for small systems. An RO unit is a common type of POU device, which is a membrane system that rejects compounds based on their molecular properties and characteristics of the reverse osmosis membrane. The RO units removed 86% of the total arsenic. These systems can remove arsenic and fluoride from the drinking water, supply of a small rural community of approximately 200 people. Ion exchange resins and adsorbent materials (activated alumina, granular activated carbon, other adsorbents) are considered acceptable for Point-of-Use (POU) devices, and smaller communities may be better able to afford POU treatment systems over central treatment.

KEY POINTS

Metal removal involves far more than removing ions from water. It also requires disposal of waste solids (e.g. spent filter material, coagulant sludge) and residual fluids likely to contain high levels of metals and possibly other hazardous or radioactive (e.g. uranium, radon) constituents. If the water requires pretreatment (e.g. lowering of pH for adsorption processes, or raising it for lime-softening) the waste situation becomes more complicated. At the same time, most treatments tend to affect effluent levels of dissolved components other than metals – sometimes in a deleterious fashion. For example, removal of arsenic through ion exchange also tends to lower bicarbonate levels increasing the corrosivity of the effluent. Coagulation using Al(III) or Fe(III) salts can result in increased concentrations of these metals in the treated water, as well as the anions (e.g. sulfate) from the salts used in the process (Krumhansl, 2009).

REFERENCES

Akpor O. B. and Muchie M. (2010). Remediation of heavy metals in drinking water and wastewater treatment systems: processes and applications. *Int. J. Phys. Sci.*, **5**, 1807–1817.

Athanasiadis K. and Helmreich B. (2005). Influence of chemical conditioning on the ion exchange capacity and on kinetic of zinc uptake by clinoptilolite. *Water Res.*, **39**, 1527–1532.

Berber-Mendoza M. S., Leyva-Ramos R., Alonso-Davila P., Fuentes-Rubio L. and Guerrero- Coronado R. M. (2006). Comparison of isotherms for the ion exchange of Pb (II) from aqueous solution onto homoionic clinoptilolite. *J. Colloid Interface Sci.*, **301**, 40–45.

Chang Q., Zhang M. and Wang J. X. (2009). Removal of Cu^{2+} and turbidity from wastewater by mercaptoacetyl chitosan. *J. Hazard. Mater.*, **169**, 621–625.

Dwivedi C. P., Sahu J. N., Mohanty C. R., Mohan B. R. and Meikap B. C. (2008). Column performance of granular activated carbon packed bed for Pb(II) removal. *J. Hazard. Mater.*, **156**, 596–603.

Driehaus W. and Dupont F. (2005). Arsenic removal - solutions for a world wide health problem using iron based adsorbents. *European Journal of Water Quality*, **36**(2), 119–132, http//dx.doi.org/10.1051/water/20053602119.

Fields K., Chen A. and Wang L. (2000). Arsenic Removal from Drinking Water by Iron Removal Plants, National Risk Management Research Laboratory, Office of Research and Development, USEPA/600/R-00/086, USA.

Fu F. and Wang Q. (2011). Removal of heavy metal ions from wastewaters: a review. *J. Environ. Manage.*, **92**, 407–418.

Gong R., Ding Y., Li M., Yang C., Liu H. and Sun Y. (2005). Utilization of powdered peanut hull as biosorbent for removal of anionic dyes from aqueous solution. *Dyes Pigments*, **64**, 187–192.

Gode F. and Pehlivan E. (2006). Removal of chromium (III) from aqueous solutions using Lewatit S 100: the effect of pH, time, metal concentration and temperature. *J. Hazard. Mater.*, **136**, 330–337.

Han I., Schlautman M. A. and Batchelor B. (2000). Removal of hexavalent chromium from groundwater by granular activated carbon. *Water Environ. Res.*, **72**, 29–39.

Hashim M. A., Mukhopadhyay S., Sahu J. N. and Sengupta B. (2011). Remediation technologies for heavy metal contaminated groundwater. *J. Environ. Manage.*, **92**, 2355–2388.

Holm T. R. and Wilson S. D. (2005). Chemical Oxidation for Arsenic Removal. A Midwest Technology Assistant Center Publication, TR 06-05, ISWS Contract Report (2006-10), IL, USA.

Johnson P. D., Girinathannair P., Ohlinger K. N., Ritchie S., Teuber L. and Kirby J. (2008). Enhanced removal of heavy metals in primary treatment using coagulation and flocculation. *Water Environ. Res.*, **80**, 472–479.

Kardos M. (2006). Arsenic removal from drinking water using granular ferric hydroxide (GFH). Proceedings of the Scientific Student Conference, Budapest University of Technology and Economics, Budapest, HU.

Khandaker N. R., Brady P. V. and Krumhansl J. L. (2009). Arsenic Removal From Drinking Water: A Handbook For Communities. Sandia National Laboratories, Albuquerque, NM, USA, 48, Chapter 4, 19–31.

Laky D. (2009). Arzénmentesítés koagulációval, doktori értekezés tézisei, Budapesti Müszaki és Gazdaságtudományi Egyetem (Arsenic Removal by Coagulation, Theses of the PhD dissertation, Budapest University of Technology and Economics). Budapest, HU, pp. 1–16.

Laky D. and Licskó I. (2011). Arsenic removal by ferric-chloride coagulation – effect of phosphate, bicarbonate and silicate. *Water Sci Technol*, **64**(5), 1046–1055. doi: 10.2166/wst.2011.419.

Lorenzen L., van Deventer J. S. J. and Landi W. M. (1995). Factors affecting the mechanism of the adsorption of arsenic species on activated carbon. *Minerals Eng.*, **8**, 557–569.

Mohan D. and Pittman J. C. U. (2007). Arsenic removal from water/wastewater using adsorbentsea critical review. *J. Hazard. Mater.*, **142**, 1–53.

Murthy Z. V. P. and Chaudhari L. B. (2008). Application of nanofiltration for the rejection of nickel ions from aqueous solutions and estimation of membrane transport parameters. *J. Hazard. Mater.*, **160**, 70–77.

Öllős G. (1998). *Víztisztítás-Üzemeltetés Handbook* (Water treatment – Operation Handbook), Egri Nyomda Ltd., Eger, HU, Chapter 13, **965**, 784–827.

Pontius F. W. (1990). Water Quality and Treatment, A Handbook of Community Water Supplies. 4th edn, American Water Works Association, McGraw-Hill, New York, USA, 269–365.

Rajakovic L. V. (1992). The sorption of arsenic onto activated carbon impregnated with metallic silver and copper. *Sep. Sci. Technol.*, **27**, 1423–1433.

Remembrance L. N. and Möller G. (2006). Arsenic Removal from Drinking Water: A Review. *Environment*, **3**(1), 133–139. http://www.blueh2o.net/docs/asreview%20080305.pdf.

Selecky M., White B. and Grunenfelder G. (2005). Arsenic treatment for small water systems, department of health, division of drinking water. *Guid. Doc.*, **46**, 13–23.

Tor A., Buyukerkek T., Cengeloglu Y. and Ersoz M. (2004). Simultaneous recovery of Cr(III) and Cr(VI) from the aqueous phase with ion exchange membranes. *Desalination*, **171**, 233–241.

Chapter 3

Oxidation for metal removal

Larry Russel, Todd Russell and Brian Croll

3.1 INTRODUCTION

There are a variety of metals in drinking water that will respond to aeration by precipitation and removal, for instance: iron, manganese, chromium III, selenium and arsenic III. Changing the oxidation state of a metal can dramatically alter the chemical properties of the metal. The changes in these chemical properties can be taken advantage of to remove metals as part of the treatment process. Generally oxidation has been used as an effective treatment technique for the removal of arsenic, iron and manganese. Discussion of the role of oxidation in the removal of arsenic is discussed in Chapter 2. Iron and manganese are by far the most common metals of concern in drinking water and the removal of these metals is the main focus of this chapter.

Iron and manganese present similar challenges when found in solution and often appear in tandem. Both metals tend to be found in solution under anoxic conditions such as groundwater or hypolimnetic layers of reservoirs and lakes. Furthermore, iron and manganese are like chemicals sharing near identical electron configurations. Both elements tend to be found as divalent ions in solution under reducing conditions, but can also be found in higher oxidation forms such as Fe(III) and Mn(IV). Both elements pose a persistent challenge to water quality. Iron is the fourth most abundant element in the earth's crust and can enter solution either through contact with soluble ferrous sulfides and oxides or via corrosion of iron and steel mains (generally as a suspended solid) (AWWA, 2003; MWH, 2005). Manganese is less prevalent than iron in natural waters, and is only a trace element in the earth's crust. It appears naturally in soils and rock as oxides, carbonates, and hydroxides, which all have sparing solubility.

Iron and manganese are non-toxic at naturally occurring levels. Iron is typically seen at concentrations no higher than 10 mg/L in water and manganese rarely exceeds concentrations of 1 mg/L (AWWA, 2003). Despite having no direct health consequences, the presence of these elements affects the quality of drinking water. Concentrations of iron and manganese above 0.1 mg/L and 0.02 mg/L respectively may impact the palatability of drinking water by imparting a metallic taste, and in the case of iron will turn a whisky and water drink black. Furthermore, the presence of iron and manganese ions can cause discoloration of laundry and plumbing fixtures as the ions precipitate out of solution. The presence of both these elements can also lead to bacterial growth of organisms such as Crenothrix and Gallionella and can cause fouling in pipes (HDR, 2001). The United States Environmental Protection Agency (USEPA) set the secondary maximum contaminant levels, SMCLs, for iron and manganese at 0.3 mg/L and 0.05 mg/L and the EU sets the standards at 0.2 and 0.05 mg/L, respectively (USEPA, 1992). The American Water Workers Association (AWWA) suggests even more stringent limits of 0.05 mg/L of iron and 0.01 mg/L of manganese.

Chemical oxidation is one process that can be used to remove dissolved iron and manganese from drinking water. The premise of chemical oxidation is to convert dissolved Fe(II) and Mn(II) to insoluble ferric hydroxide and magnesium oxides. The precipitate can then be extracted from solution by sedimentation or filtration. The common oxidants used for iron and manganese removal are oxygen, chlorine, chlorine dioxide, ozone and potassium permanganate. The effectiveness of oxidation on iron and manganese hinges on several key factors: the pH of the solution, the reductive potential of the solution, the presence of naturally occurring organic material (NOM), and the ionic strength of the solution.

Iron and manganese do have some distinct traits that uniquely affect how each element is oxidized in solution. While both iron and manganese can be removed from solution by oxidation, Mn(II) is substantially more stable than Fe(II) in solution which typically leads it to require higher detention times (AWWA, 2003). Manganese is also autocatalytic where iron is not. As MnO_2 is generated in solution it will adsorb dissolved Mn(II). Not only does this adsorption provide a catalyst for the reaction of Mn(II) to MnO_2, it also provides a second pathway, adsorption, for Mn(II) to be removed from solution (MWH, 2005). Unlike manganese, iron will tend to form complexes with NOM such as humic and tannic acids. These complexes, also known as filterable iron, dramatically reduce the effectiveness of oxidation by oxygen, chlorine, chlorine dioxide, and potassium permanganate (MWH, 2005). At pH values commonly associated with natural water, pH 5 to 9, the removal of filterable iron commonly requires additional treatment steps, such as, alum coagulations or carbon filtering.

Due to the complex chemistry involved in the removal of Fe(II) and Mn(II) with oxidation, design and testing on the bench and pilot scale are strongly encouraged. Many systems have been developed with only a limited understanding of the chemistry and timescales required, and have failed because of these shortcomings. Insufficient detention times have been a common pitfall for designers, which must be avoided for effective removal of these metals with oxidation.

3.2 OVERVIEW OF IRON AND MANGANESE REMOVAL

Although Fe(II) and Mn(II) present similar challenges and are often found in tandem, the removal of each of these metals requires understanding of their different chemistries and removal behaviors. The following sections are divided so as discuss the important factors to achieve effective removal of iron and manganese by oxidation.

3.3 IRON REMOVAL OVERVIEW

Ferrous iron [Fe(II)] can be removed with a variety of oxidants including oxygen, chlorine, chlorine dioxide, ozone and potassium permanganate. Each process has different pros and cons and the source water characteristics greatly influence the feasibility of each oxidant.

The presence of NOM complexes with Fe(II) greatly reduces the reaction rates for chemical oxidation and can render it entirely ineffective. There is limited understanding of the complicated chemistry involved in iron complexes with NOM. Recognition of the potential for the presence of these complexes in source water is an important factor when considering the feasibility of Fe(II) removal with oxidation.

3.4 IRON REMOVAL VIA AERATION

Aeration can be used to provide oxygen to oxidize Fe(II) to form precipitates. Using aeration is inexpensive and advantageous, because it can be implemented without additional chemical storage and handling concerns. However, its slower reaction rates can make aeration impractical in certain waters. Typical design of aeration systems include air-diffusion-type or coke tray aerators followed by a detention basin terminating in a sand or charcoal filter (JMM, 1985; MWH, 2005). Aeration is typically operated at an air-water ratio of 0.75:1.0 and the aeration is typically applied at a depth of 3–5 m with an oxygen transfer efficiency of 5–10% (JMM, 1985). Following aeration, typical detention times range from 15–30 minutes (MWH, 2005).

Coke tray aerators have been used for aeration, due to their oxygen transfer efficiency. Design of these aerators typically consists of 3–5 successive perforated stainless steel trays containing crushed coke or other high surface area material. These aerators increase oxygen transfer and are quickly coated in iron oxides once in operation. This oxide coating assists in the oxidation process. Typical water loading rates for these trays range from 600 to 800 $L/m^2 * min$ (15 to 20 gpm/ft^2) (MWH, 2005).

Stoichiometry for the oxidation reaction of Fe(II) with oxygen is (Faust & Aly, 1998):

$$O_{2(g)} + 4Fe^{2+} + 10H_2O = 4Fe(OH)_{3(s)} + 8H^+$$

Reaction kinetics for iron oxidation with aeration are believed to be rate limited by the reaction of $O_{2(g)}$ with Fe(II). The rate law for this reaction is first order with respect to both $[O_{2(g)}]$ and [Fe(II)]. The rate law for pHs above 4.5, with alkalinity of $2.9–3.9 * 10^{-2}$ eq/L (140–200 mg/l as $CaCO_3$) is (Faust & Aly, 1998):

$$-d[Fe(II)]/dt = k[OH^-]^2 P_{O_2}[Fe(II)]$$

where

[Fe(II)] = total concentration of ferrous iron (M)

k = rate constant (M^{-2}-atm^{-1}-min^{-1})

P_{O_2} = partial pressure of oxygen

In practice however, the kinetics are much more complicated than indicated by the above rate law. The kinetics are influenced by a number of factors including pH, iron complexes with NOM, ionic strength, alkalinity and temperature. A study of iron removal from Illinois ground waters with aeration found $t_{1/2}$ values of 4.3–54 minutes with a detention time of 60 minutes for aeration (Faust & Aly, 1998). These slow reaction half lives indicate that historically iron removal systems utilizing aeration have not been designed with sufficient aeration and detention times.

The kinetics of oxidation via aeration are also heavily influenced by pH. Aeration is generally ineffective at a pH less than 5, and the reaction rate increases 100-fold for every unit increase in pH above 5. In waters with low pH, pH manipulation is an important key for effective iron removal with aeration. Increasing pH for metal removal may require neutralization later in the treatment train. These cost tradeoffs should be carefully considered for an aeration system. Aeration typically requires a higher pH to be effective than strong chemical oxidants such as chlorine.

When considering aeration for iron removal it is important to perform pilot tests to assess the kinetics of Fe(II) oxidation. The required aeration and detention times vary widely with different water characteristics and there is only limited ability to correctly predict the required times. In many cases, particularly when iron complexes are present, aeration will be impractical for iron removal because of the long detention times required.

3.5 IRON REMOVAL VIA CHLORINATION

Iron removal via oxidation with chlorination followed by filtration can be a cost effective option, but care must be taken to avoid negative side effects, such as, disinfection byproducts. Traditionally free chlorine doses as high as 5 mg/L were used to effectively oxidize Fe(II). However, regulatory changes to increasingly stringent control of disinfection byproducts (DBP) have made oxidation with chlorine less attractive (MWH, 2005). If DBP are not a concern, chlorine may be an attractive choice when a treatment plant has existing infrastructure for chlorine disinfection as it is possible for the oxidation chlorine dose to work as part of the disinfection credit.

Chlorine is a strong oxidant and can effectively oxidize Fe(II) to form iron oxides which can be removed by precipitation. Typical design for removal of iron via chlorination consists of a well mixed chemical injection basin followed by a detention basin and finally direct filtration. Removal takes place by the formation of oxide precipitates. In some cases, the addition of alum coagulation and permanganate conditioned filter media can be implemented to reduce chlorine doses and the associated DBP.

Typically chlorine dosing to achieve a free chlorine residual of approximately 0.4 mg/L is required for effective oxidation of iron (JMM, 1985). This reaction has been shown to be sensitive to pH and is most effective at a pH above 5. Typically chlorine oxidation can be effective at a lower pH than is required when using aeration. To achieve oxidation times of between 15 and 30 minutes a pH of between 8 and 8.5 is required (MWH, 2005). In waters containing ammonia, the formation of chloramines will greatly reduce the oxidation of Fe(II).

The stoichiometry for the oxidation reaction of Fe(II) with chlorine is (Faust & Aly, 1998):

$$2Fe^{2+} + 6H_2O + Cl_{2(g)} = 2Fe(OH)_{3(s)} + 2Cl^- + 6H^+$$

Chlorine oxidation has an added benefit of rapid kinetics which allow for much shorter detention time when compared to aeration. The shorter detention times compared with aeration can be more pronounced when influent iron has formed complexes with organic substances (JMM, 1985). However, a variety of more recent studies have shown that iron complexes with certain natural humic acids can make chlorine oxidation of Fe(II) ineffective (Faust & Aly, 1998). The presence of dissolved silica in source waters can also increase the detention time required for oxidation by chlorine, but literature on this trend is conflicting (Faust & Aly, 1998).

3.6 IRON REMOVAL VIA CHLORINE DIOXIDE

Chlorine dioxide is a stronger oxidant than chlorine and has been shown effective for Fe(II) removal via oxidation. System designs are similar to those used in chlorine oxidation, except for the initial chemical dosing stage. In the absence of organic matter, chlorine dioxide reacts rapidly to oxidize iron. This reaction is also effective at cold temperatures (Faust & Aly, 1998). Generally the oxidation reactions with iron and chlorine dioxide happen very rapidly, particularly at a pH above 5.5 (MWH, 2005). Fe(II) complexes with NOM are highly resistant to oxidation with chlorine dioxide.

The stoichiometry for oxidation reaction of Fe(II) with chlorine dioxide is (Faust & Aly, 1998):

$$ClO_{2(g)} + 5Fe^{2+} + 13H_2O = 5Fe(OH)_{3(s)} + 11H^+ + Cl^-$$

3.7 IRON REMOVAL VIA OZONE

Ozone can be utilized as an oxidant for the removal of Fe(II), but has rarely been implemented for this purpose in the US. A comparison study of aeration and ozonation for removal of ferrous iron in groundwater found that the kinetics are much faster for ozone, and that ferrous iron is removed more completely when ozone was used (Cromley & O'Connor, 1976).

Ozone has advantageous over the use of chlorine as it does not have the same issues with disinfection byproducts. European treatment plants with preozonation have shown effective removal of Fe(II) (MWH, 2005).

The stoichiometry for the oxidation reaction of Fe(II) with ozone is (Faust & Aly, 1998):

$$2Fe^{2+} + 5H_2O + O_{3(g)} = 2Fe(OH)_{3(s)} + 4H^+ + O_{2(g)}$$

One caveat in the use of ozone oxidation for waters that contain iron and manganese is that overdose of ozone can oxidize Mn(II) to Mn(VII) (permanganate). In this process the permanganate formed turns the water pink and is easily detected by consumers.

3.8 IRON REMOVAL WITH POTASSIUM PERMANGANATE ($KMNO_4$)

Potassium permanganate ($KMnO_4$) has been demonstrated as an effective strong oxidant for the removal of Fe(II). Typical design consists of a chemical dosing stage followed by a detention basin terminating in filtration through specially prepared media. The filtration method commonly used is pressure filtration with loading rates of 240–480 m/d (4–8 gpm/ft^2) (MWH, 2005). The filter media often consists of a manganese greensand filter, but can also consist of silica and/or anthracite coal (JMM, 1985; Faust & Aly, 1998). Generally $KMnO_4$ has very rapid reaction kinetics, and removal is catalyzed by the filter media. $KMnO_4$ oxidation is most effective at pH values over 5.5 (MWH, 2005). Typical detention times after chemical feed are 5–10 minutes, and can be influenced by water temperature (JMM, 1985). Although potassium permanganate tends to be more expensive than chlorine or ozone, on a per-oxidation basis, it can be efficient at iron removal and may costs less due to lower equipment and capital investment. If manganese greensand is used as the filter media, the media will require periodical recharging with potassium permanganate.

Potassium permanganate is subject to similarly reduced reaction rates in the presence of NOM complexes with Fe(II). While potassium permanganate has been shown to remove iron complexes with fulvic acid, the reaction requires contact times in excess of one hour which likely limits the feasibility of its use in this scenario (Faust & Aly, 1998).

The stoichiometry for the oxidation reaction of Fe(II) with potassium permanganate is (Faust & Aly, 1998):

$$3Fe^{2+} + 7H_2O + MnO_4^- = 3Fe(OH)_{3(s)} + MnO_{2(s)} + 5H^+$$

Care must be taken avoid overdose of potassium permanganate. An overdose in the range of 0.05 mg/L results in an easily detectable pink color (MWH, 2005).

Permanganate coated filter media can also be used in conjunction with chlorination. This application allows for reduced chlorine doses, which can thereby reduce DBP formation (MWH, 2005) and is discussed further in the sections on manganese removal.

3.9 MANGANESE REMOVAL OVERVIEW

Mn(II) can be removed via oxidation with a variety of oxidants including oxygen, chlorine, chlorine dioxide, ozone and potassium permanganate. Each process has different pros and cons and the source water characteristics greatly influence their feasibility. The oxidation of manganese is more complex than that of iron and is not as well understood. Because of this decreased understanding, designing systems for the removal of manganese with oxidation can be challenging. Factors including complex chemistry, slow reaction rates, multiple oxidation states and complexes all must be taken into account. Typical system designs are similar to those described previously for iron removal. Instances where design or operation differs from those described in the iron sections are noted in the following sections.

3.10 FILTRATION OVERVIEW

For most systems utilizing oxidation for removal of iron and manganese, proper selection of the filter media and design of the filtration unit is important. Commonly precipitation of iron and manganese will not occur until the water meets the filtration media. In the filter, previously precipitated oxides act as a catalyst to precipitate oxides out of solution (MWH, 2005). In the case of the autocatalytic reaction that that takes place in the filter is critical to the removal of manganese in these systems. While the filter plays an integral role in the precipitation, it is rare to observe excessive build up of oxides on the filter media. However, in waters containing high concentrations of iron and manganese, suspended solid accumulation in the filter can be a concern (JMM, 1985). Proper selection of filtration media is important. Filtration media typically have effective sizes >1.5 mm and low uniformity coefficients. Coal is a good media choice for this application due the ability to backwash with lower velocities, but other media such as sand can be used (JMM, 1985). Typical loading rates range from 300–600 m/d (5–10 gpm/ft^2). When implemented correctly, filters will acclimate within three weeks and produce high quality effluent with low iron and manganese (JMM, 1985).

The use of potassium permanganate as an oxidant requires different filter media and design considerations. The filters usually consist of pressure filters with smaller effective sized media that is precondition with permanganate. These filters typically run at higher loading rates of 240–480 m/d (4–8 gpm/ft^2) (MWH, 2005).

3.11 MANGANESE REMOVAL VIA AERATION

Aeration can be used to supply oxygen for the oxidation of Mn(II) to form manganese oxides. These oxides can then be removed by a combination of detention time and filtration. Typical design is similar to that used for iron removal via aeration, but requires longer detention times and higher oxygen doses. In addition, the filter plays a more critical role in the precipitation of magnesium oxides than in iron oxides.

In manganese removal via aeration it is common that the majority of material removal takes place in the filter. This is the case because of the autocatalytic nature of manganese oxides, which adsorb Mn(II) and accelerate the further formation of manganese oxides. Therefore, proper filter media and break-in are an important component of manganese removal. Proper filter break-in is necessary to build up sufficient manganese oxides on the filter media so as to allow for effective removal of Mn(II) via the autocatalytic reaction. This autocatalytic reaction is often observed by water appearing clear in the detention basin after aeration, but still containing microscopic colloidal manganese which precipitates once contacting the filter media (JMM, 1985).

The stoichiometry for the oxidation reaction of Mn(II) with oxygen is (Faust & Aly, 1998):

$$O_{2(g)} + 2Mn^{2+} + 2H_2O = 2MnO_{2(s)} + 4H^+$$

However, the autocatalytic nature of manganese oxidation means that removal rarely takes place in these simplified stoichiometric ratios.

Oxidation of manganese via aeration is considered ineffective below a pH of 9.5, and is slow, requiring a 1 hour detention time for oxidation of Mn(II) (MWH, 2005). pH is an important consideration for manganese removal and pH adjustment should be considered as part of the treatment costs.

A rate law for manganese oxidation by oxygen is (Faust & Aly, 1998):

$$-d[Mn(II)]/dt = k_1[Mn(II)] + k_2[Mn(II)][MnO_{2(s)}]$$

where k_2 and k_3 are related such that: $k_2 = k_3[OH^-]^2 P_{O_2}$

This rate law arises from three reactions involved in forming manganese oxides. As noted earlier, the chemistry of manganese oxidation is significantly more complex than that of iron, and is often poorly understood by system designers. The autocatalytic nature of manganese also complicates understanding reaction rates and the required oxidant dose. Overall, the long detention time required with aeration to achieve oxidation of Mn(II) make it impractical for manganese removal in most treatment plants.

3.12 MANGANESE REMOVAL VIA CHLORINE

The strong oxidizing characteristics of chlorine can be used to oxidize Mn(II) for removal. While oxidation of Mn(II) can occur relatively quickly, it has colloidal properties which result in much slower flocculation and precipitation than occurs with iron oxidation. These slower rates introduce some logistical issues, in addition to the DBP issue associated with higher chlorine doses that are sometimes required. To overcome these slower rates, the addition of a coagulant can assist in the formation of manganese oxide precipitates.

For comparison, at a pH of 8.0 to 8.5, chlorine can be used to oxidize Fe(II) in 15–30 minutes. Under the same conditions to oxidize Mn(II) requires 2–3 hours (MWH, 2005). Generally the rate increases with pH. Typically a 0.5 ppm free chlorine residual throughout the plant is required to remove Mn(II) (Faust & Aly, 1998). Chloramines have very little impact on Mn(II) concentrations.

The stoichiometry for the oxidation reaction of Mn(II) with chlorine is (Faust & Aly, 1998):

$$Mn^{2+} + 2H_2O + Cl_{2(g)} = MnO_{2(s)} + 4H^+ + 2Cl^-$$

To decrease DBP and increase Mn(II) removal, a hybrid system consisting of prechlorination, alum coagulation, sedimentation and filtration can be used. The filtration media is conditioned with permanganate to generate a manganese oxide coating on the surface. This coating catalyzes the removal of Mn(II). By adjusting the chlorine dose with the Mn(II) concentrations, the system can be operated in a manner that produces much fewer DBP and still removes Mn(II) effectively. In this hybrid system source waters that contain both iron and manganese can also be effectively treated. The prechlorination dose can be controlled to oxidize the iron and limit DBP formation.

3.13 MANGANESE REMOVAL VIA CHLORINE DIOXIDE

The strong oxidant properties of chlorine dioxide effectively oxidize Mn(II). Chlorine dioxide is a good alternative to chlorination, because it produces less DPB and has faster reaction rates for the oxidation of Mn(II). With water at a pH above 5.5 and without NOM, complete oxidation of Mn(II) can occur in 20 seconds (MWH, 2005). Generally the presence of NOM does not impact the oxidation rates of Mn(II). Typically the dose of chlorine dioxide for Mn(II) oxidation is 10 times that required for Fe(II) oxidation (MWH, 2005).

The stoichiometry for the oxidation reaction of Mn(II) with chlorine dioxide is (Faust & Aly, 1998):

$$5Mn^{2+} + 6H_2O + 2ClO_{2(g)} = 5MnO_{2(s)} + 12H^+ + 2Cl^-$$

3.14 MANGANESE REMOVAL VIA OZONE

Ozone has not be used commonly in the US for Mn(II) removal, but has been found to be effective for this purpose in treatment plants in Europe that utilize preozonation followed by typical treatment processes. The stoichiometry requirements for oxidation of Mn(II) are 0.87 mg O_3/mg Mn^{2+} (MWH, 2005). Overdose of ozone in water with manganese can lead to the formation of permanganate which leaves the water with an easily detectable pink color.

3.15 MANGANESE REMOVAL VIA POTASSIUM PERMANGANATE ($KMNO_4$)

The very strong oxidant properties of potassium permanganate ($KMnO_4$) can be used to rapidly oxidize Mn (II) over a wide range of pHs. At a pH of 5.5 with NOM present Mn(II) can be oxidized in $<$20 seconds (MWH, 2005), which is much faster than other oxidants. However, the higher costs associated with potassium permanganate oxidation have resulted in less use compared to other oxidants such as chlorine. A common practice in situations where both Mn(II) and Fe(II) are present is to first oxidize the ferrous iron with chlorine, then add $KMnO_4$ to oxidize the manganese. This process is usually combined with filtration using specially conditioned filter media. This combined process performs best at pH above 7.5 (MWH, 2005). Care must be taken avoid overdoses of potassium permanganate. An overdose in the range of 0.05-mg/L results in an easily detectable pink color (MWH, 2005).

3.16 REMOVAL OF IRON AND MANGANESE USING MICROBIOLOGICALLY ACTIVE FILTERS (BIOLOGICAL IRON AND MANGANESE REMOVAL)

Under the appropriate water quality conditions naturally occurring bacteria in anaerobic waters will colonise filter beds. The resulting microbiologically active filters will remove iron and manganese from the source water at very high flow rates (up to 50 m/h) (Water Treatment Handbook) giving highly cost-effective treatment. The process is applicable to most anaerobic groundwaters and many plants are in operation in N Europe.

Iron

The iron must be in the ferrous state as it reaches the filter, biological processes will not remove ferric iron. Before filtration the dissolved oxygen is raised to 5 to 10% (typically 8%) saturation by simple air contact in the top of an open filter, air injection or recirculation of the water on top of the filter. The pH may need adjustment. The filters are typically packed with sand of 0.9 mm average diameter (8/16 mesh) to a depth of 1.5 to 2 m but other configurations are possible. Biological activity normally develops within 3 days and once established the product water will be below 30 μg/l iron at high filtration rates provided that the input iron concentration is not too high ($>$5 mg/l). If the iron concentration is very high then a second stage of filtration may be needed, either microbiological or chemical depending on the dissolved oxygen concentration of the filtered water and the valency state of the iron. Filtration rates of up to 50 m/hr are possible.

Shallower filter beds (0.6 to 1 m deep) operated at high rates (up to 50 m/h) may be used as an alternative to settlement where iron concentrations need to be reduced prior to a physico-chemical iron and manganese removal plant.

Backwash of the filters is by separate or simultaneous air and water, both have given satisfactory results. The volumes of the backwash water and backwash solids are much less per mg/l of iron removal than for physico-chemical iron removal.

Manganese

As with iron, the manganese must be in the reduced form but, unlike iron, the dissolved oxygen must be raised to at least 30% saturation prior to filtration and the process takes several weeks to establish. Treatment rates of up to 40 m/h have been quoted (Water Treatment Handbook). Ammonia is often present together with iron and manganese in groundwaters. Ammonia can be removed microbiologically on the same filter as manganese but treatment rates are lower than for manganese at N European groundwater temperatures (UK 11°C), leading to plants typically being operated at about 10 m/h for the dual role.

KEY POINTS

A range of metals in drinking water (iron, manganese, chromium III, selenium, arsenic III) can be removed by aeration and precipitation. However, by far the most common metals of concern in drinking water are iron and manganese. Both elements tend to be found in solution under anoxic conditions such as groundwater or hypolimnetic layers of reservoirs and lakes. Both elements tend to be found as divalent ions in solution under reducing conditions, but can also be found in higher oxidation forms such as Fe(III) and Mn(IV).

Chemical oxidation can be used to remove dissolved iron and manganese from drinking water. Chemical oxidation converts dissolved Fe(II) and Mn(II) to insoluble ferric hydroxide and magnesium oxides and the precipitates can then be extracted from solution by sedimentation or filtration. The common oxidants used for iron and manganese removal are oxygen, chlorine, chlorine dioxide, ozone and potassium permanganate. The effectiveness of oxidation on iron and manganese hinges on several key factors: the pH of the solution, the reductive potential of the solution, the presence of naturally occurring organic material (NOM), and the ionic strength of the solution. Mn(II) is substantially more stable than Fe(II) in solution and manganese is autocatalytic where iron is not. Due to the complex chemistry involved in the removal of Fe(II) and Mn(II) with oxidation, design and testing on the bench and pilot scale are strongly encouraged.

REFERENCES

AWWA (2003). Water Quality: Principles and Practices of Water Supply Operations. American Water Works Association, Denver.

Cromley J. T. and O'Connor J. T. (1976). Effect of ozonation on the removal of iron from a ground water containing organic substances. *J. Am. Water Works Assoc.*, **68**, 315.

Faust S. D. and Aly E. M. (1998). Chemistry of Water Treatment. 2nd edn, Ann Arbor Press Inc., Chelsea, Michigan.

HDR Engineering, Inc (2001). Handbook of Public Water Systems. 2nd edn, John Wiley & Sons, Inc., Hoboken, New Jersey.

Montgomery J. M. Consulting Engineers Inc. (1985). Water Treatment Principles and Design. John Wiley & Sons, Inc., New York.

MWH (2005). Water Treatment: Principles and Design. 2nd edn, John Wiley & Sons, Inc., Hoboken, New Jersey.

USEPA (1992). Secondary Drinking Water Regulations: Guidance for Nuisance Chemicals. United States Environmental Protection Agency, Washington, D.C. [http://www.epa.gov/ogwdw000/consumer/2ndstandards.html][accessed June 2010].

Water Treatment Handbook (1991). Water Treatment Handbook. Degremont, Lavoisier Publishing, Paris, France, pp. 1206–1210 and 1214–1216.

Chapter 4

Coagulation, flocculation and chemical precipitation

Mehmet Emin Aydın, Zdravka Lazarova, Ali Tor and Senar Ozcan

4.1 INTRODUCTION

One of the major objectives of the water treatment is the removal of suspended matters from water. For this purpose, clarification involving coagulation, flocculation, settling, filtration is performed. The process of coagulation, along with flocculation, is employed in order to provide effective clarification by destabilization of colloidal materials. Coagulation, flocculation and clarification processes are also effective in removing metals from water. This practice guide focuses on coagulation and flocculation which are the two key steps affecting the finished water quality.

4.2 DESCRIPTION OF TECHNOLOGIES IMPLEMENTATION

Coagulation, flocculation, sedimentation, filtration and disinfection are components of conventional water treatment systems (Figure 4.1). Coagulation and flocculation are basically employed to destabilize particles and agglomerate dissolved and particulate matters. Sedimentation and filtration removes solids.

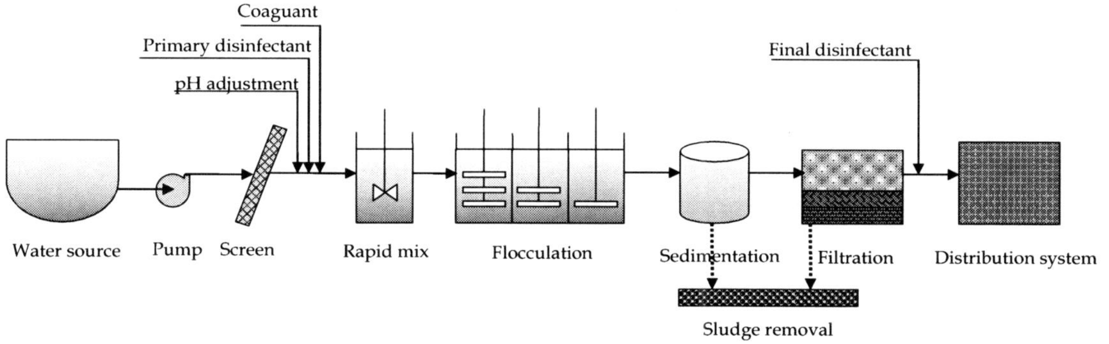

Figure 4.1 Conventional water treatment processes.

4.3 COAGULATION

Coagulation is a process in which coagulant is dispersed into the treated water by intensive agitation. The size of the particles increases due to their destabilization and formation of larger aggregates. Flocculation is an operation in which small particles agglomerate into well defined flocs by gentle mixing for a longer time. The flocs can be effectively removed in the following separation processes such as sedimentation, flotation and filtration.

Coagulants are used in water treatment mainly for destabilization of colloidal suspensions and agglomerate the particles to form settleable flocs. They are also used to form precipitates that adsorb natural organic matters and some inorganic materials such as phosphates, arsenic compounds and fluoride. Aluminum and iron based coagulants as well as organic polymers (polyelectrolytes) are widely available and are most commonly used for water treatment. The most common aluminum and iron salts utilized are: aluminum sulphate (alum), aluminum chloride, ferric sulphate and ferric chloride. Their complex forms (prehydrolyzed or pre-polymerized salts) commonly available are: poly aluminum sulphate, poly aluminum silicate sulphate, poly aluminum chloride, pre-hydrolyzed alum, hydroxylated ferric sulphate, poly ferric sulphate and poly ferric chloride (Baruth, 1990; Niquette *et al.* 2004; Tillman, 1996). Overdosing the coagulants can result in restabilization of the suspension (Melia, 1982).

When metal salts (e.g. $Al_2(SO_4)_3$ or $FeCl_3$) are added as coagulants to the water, different hydrolysis products are formed (Figure 4.2). At low pH, positively charged soluble hydrolysis products and aqua-metal ions Al^{3+} and Fe^{3+} are formed while at high pH negatively charged, soluble hydrolysis products $Al(OH)_4^-$ and $Fe(OH)_4^-$ are formed. At alkaline pH values (pH > 8), the monomeric anion $Al(OH)_4^-$ is the main soluble species; at lower pH values (pH < 6), Al^{3+} and $Al(OH)^{2+}$ are the main soluble species (Hudson, 1981; Tillman, 1996).

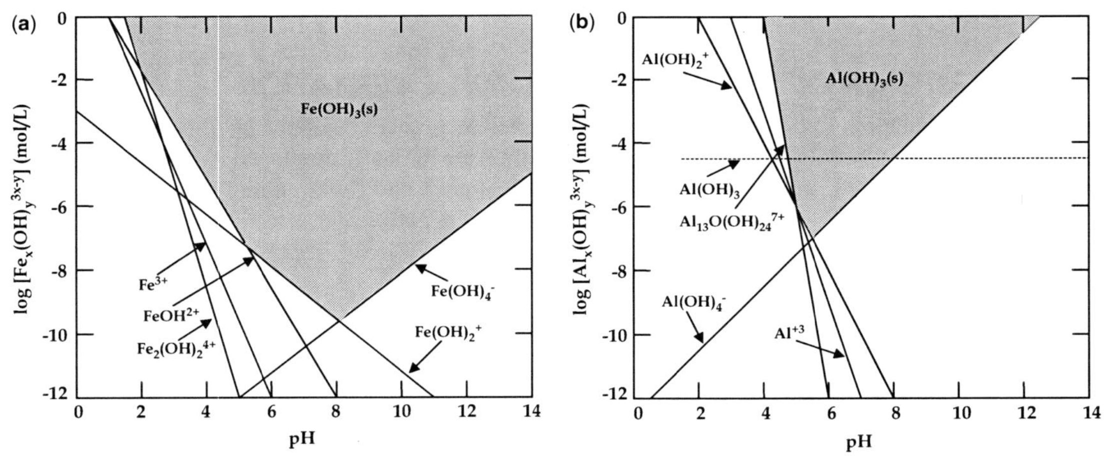

Figure 4.2 Hydrolysis of aluminum or iron salts.

When commercial alum is added to water, hydrolysis reactions occur prior to settleable solid $Al(OH_3)_{(s)}$ formation. During these reactions hydrogen ions are produced and hence the alkalinity of the water is reduced. About 0.5 mg L^{-1} alkalinity as $CaCO_3$ is consumed for each 1 mg L^{-1} alum used. When treating low alkalinity waters, lime or caustic soda is added to water in order to get necessary alkalinity for optimum coagulation.

Cationic monomers such as Fe^{3+} and $Fe(OH)_2{}^+$ are the main soluble products at low pH (pH < 6) while at high pH (pH > 10), the monomeric anion $Fe(OH)_4{}^-$ is main soluble species. Fe(III) is least soluble at about pH 8. Iron (III) is a stronger acid and less soluble than aluminum and also consumes alkalinity during precipitation (Hudson, 1981; Tillman, 1996).

Coagulant aid is used to improve coagulation process by forming heavier flocs as well as to reduce the amount of primary coagulant used. Activated silica is used to strengthen the flocs, it is generally produced by activating sodium silicate with an acid. Activated silica is added after the primary coagulant. Dose optimization is important since too much silica may reduce coagulation and clog filers.

Oxidative chemicals such as chlorine, ozone or potassium permanganate can be added to improve the coagulation by oxidizing dissolved organics. The oxidizing chemicals are added prior to the primary coagulant to convert the dissolved organics more easily.

Adjustment of the pH-value may also be considered as a coagulant aid (pH adustment shouldn't be considered as a coagulany "aid" as it a critical element of efficient coagulation) since all the coagulants have an optimum operation pH range. Outside of this range, the flocs can become soluble and pass through the subsequent sedimentation and filtration unit. Lowering the pH-value is done by adding sulfuric acid or hydrochloric acid while increasing pH is carried out by adding lime, caustic soda or soda ash. These chemicals may also be used to add necessary alkalinity if the treated water. Polyelectrolytes or polymers are used widely as other coagulant aids. The polymers are big molecules that produce charged ions or electrolytes when dissolved in water. There are three types of polyelectrolytes such as cationic, anionic or nonionic. Cationic polymers may be used alone as a primary coagulant. However, they are often employed as coagulant aid with a metallic coagulant in order to reduce the dose required. Polymer improved flocs also settle quicker. Anionic polymers form negative ions and are used to provide more reactions sites for positively charged coagulants. Use of anionic polymers may reduce the coagulant dosage to form heavy flocs and may enhance color removal. However, overdose of anionic polymers may reduce efficiency of the coagulation process. Nonionic polymers are neutral however when dissolved in water they form positive and negative ions. Nonionic polymers are used as primary coagulants, coagulant aid and filter aid. The applied dosage ranges of cationic, anionic and nonionic polymers are given in Table 4.1. Overdose of polymers may reduce efficiency of coagulation and flocculation (Baruth, 1990; Tillman, 1996; Drinan, 2000).

Table 4.1 Usually applied dosage ranges of cationic, anionic and nonionic polymers.

Type of polymer	Usually dosage range (mg L^{-1})
Cationic	0.1–1.0
Anionic	0.1–1.0
Nonionic	1.0–10

4.4 COAGULATION REACTORS

Various mixing devices are currently employed for coagulations in water treatment including back mix reactors, in-line blenders, hydraulic pumps, diffusers and injection devices, and static mixers. Widely used rapid mixing units in water treatment are mechanical back mix reactors (Figure 4.3).

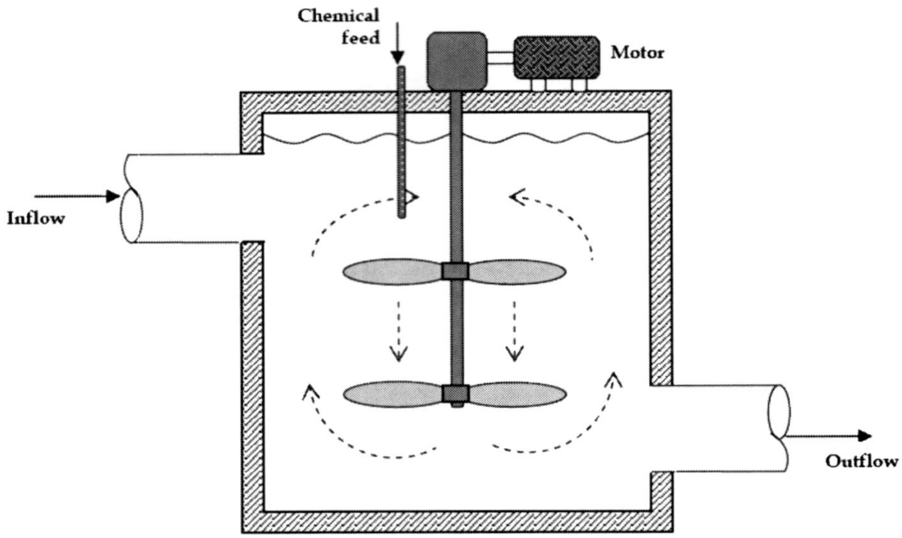

Figure 4.3 Rapid mix reactor.

Back mix reactors are designed usually with mean velocity gradients ranging from 700 to 1000 s^{-1} with contact time ranging from 1 to 2 min. In manufactured in-line blenders destabilization occur by charge neutralization within 0.01 s to 1 s. In this sort of mixers G values range from 3000 to 5000 s^{-1} with contact time of 0.5 to 1.0 s. In line-blenders provide good mixing with little short circuiting and cost could be reduced by avoiding conventional rapid mixing units. Hydraulic pumping is sometimes used for mixing coagulant in water treatment plants. The coagulants are added prior to the hydraulic pump with G values about 800 s^{-1} and residence time 2 s. Hydraulic pump mixing is advantageous since there is no mechanical part required however, flexibility in operation is limited since mixing intensity can not be changed. In diffusers and injection type devices coagulants are fed through orifices in a series of tubes or through Venturi arrangements. Velocity gradients in these units change from 700 to 1000 s^{-1} with mixing times 0.5 to 1.0 s. static mixers generate turbulence and mixing from the fixed vanes within the mixer. Mixing efficiency depend on the flow rate through the mixer (Baruth, 1990; Letterman, 1999; Pontius, 1990).

4.5 FLOCCULATION

Flocculation is carried out in a slow mixing tank after addition of the coagulant to the treated water in a rapid mixing tank Settleable flocs are formed by gentle mixing with energy gradient ranging from 20 to 70 s^{-1} for a period of between 10 and 30 minutes. If flocculation is followed by filtration, small dense flocs are required at the higher end of the energy range. However, if settling follows the slow mixing, larger dense flocs resistant to breaking up are required at the lower end of the energy range. Floc formation begins in 2 s of coagulant addition and mixing. Flocs may be broken if turbulence in the tank is high and broken flocs may not settle easily or re-form. Optimum floc that is readily settled or filtered are formed in a flocculation chamber that has gradually reduced energy. The slow mixing process in flocculation is carried out to maximize the contact of destabilized particles to form settleable or filterable flocs (Pontius, 1990; Letterman, 1999; G. Tchobanoglous, 1990; Tillman, 1996). Compartmentalization of the flocculation chamber is suggested by dividing the chamber into two or more stages as shown in Figure 4.4. Compartments will facilitate reduced tapered energy zones and will reduce short circuiting.

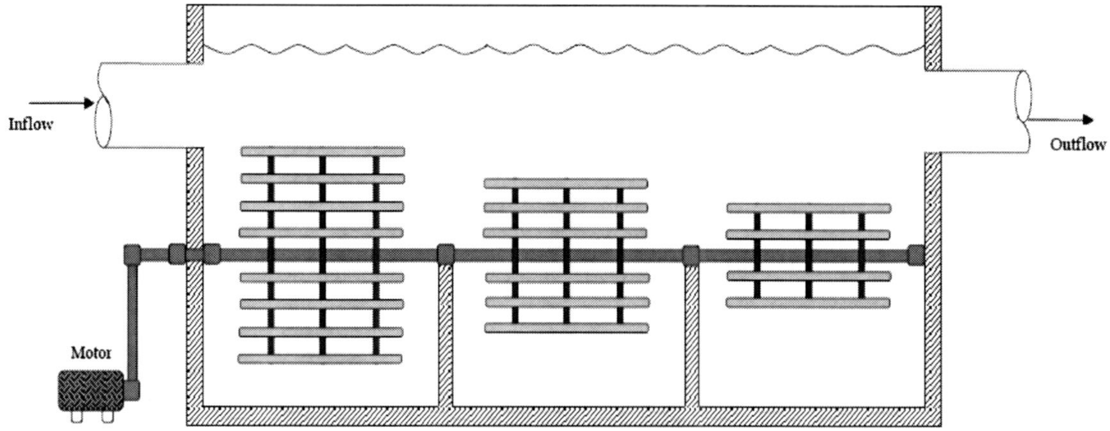

Figure 4.4 Horizontal flow flocculation tank.

4.6 FLOCCULATION REACTORS

Flocculation process occurs in a slow mixing tank (Figure 4.4). The mixing may be brought about mechanically or hydraulically. Three types of mechanical mixers such as paddle or reel-type, turbines and axial flow propellers are generally employed in flocculation units. These mixers may be installed horizontally or vertically. Paddle or reel type mixers rotate at 2–15 rpm and their tip speeds change between 0.3–0.7 m/s. Plate turbine flocculators are made of flat blades connected plane of a rotating shaft which has speeds about to 10–15 rpm. An Axial flow propeller has blades inclined at 35°C to the plane perpendicular to the rotation axis. Axial flow propeller may be installed horizontally or vertically and may be operated between 150–1500 rpm up to velocity gradients of 90 s^{-1} improved efficiency in flocculation may be obtained employing three or four compartments with tapered velocity gradients from 40–60 s^{-1} in the first compartment down to 15–25 s^{-1} in the last compartment.

Hydraulic mixing by means of baffles is employed for horizontal flow flocculators. Hydraulic head loss due to baffles is associated with velocity gradient. Baffle flocculators are usefull in reducing mechanical equipment. Solid contact clarifiers combine the processes of flocculation and up-flow clarification. Construction of solid contact clarifier is economical compared with separate horizontal flow units (Pontius, 1990).

Jar testing is an important and widely accepted tool for coagulation flocculation. The jar test is used for determination of various parameters of coagulation-flocculation process such as coagulant and dosage selection, coagulant aid and dosage selection, determination of optimum pH, determination of chemical addition point, determination of intensity and duration of mixing for rapid and flow mixing. Laboratory scale jar test apparatus consist of several jars and mixing paddles. Dosage optimization by jar test usually include following steps: different coagulant doseses are added to the jars near the mixing paddles while rapid mixing the same water. Rapid mixing is carried out for 0.5 to 1.0 minute at the maximum speed usually at 100 rpm. The suspension then mixed slowly for 15 to 20 min at speed between 25–35 rpm.

The suspension is let stand still to allow settlement of flocs for 30 to 45 minutes without mixing. Turbidity or any parameter that is expected to be reduced is measured of the samples taken from supernatant of the jars. The lowest residual turbidity or other measured parameter corresponds to the optimum coagulant dose (Letterman, 1999; Hudson 1981).

4.7 CHEMICAL PRECIPITATION

Chemical precipitation is employed in water treatment as an effective process such as coagulation with alum, ferric sulfate, ferrous sulfate and lime softening. Efficiency of chemical precipitation depends on the solubility of the complexes formed. Heavy metals existing as cations in water form insoluble hydroxides and carbonates and precipitate. This is achieved usually by adding caustic soda or lime to adjust the pH according to the maximum metal insolubility.

In Table 4.2, the efficiency of metal removal from drinking water by coagulation and lime softening is presented.

Table 4.2 Removal of metals by coagulation and lime softening.

Contaminant	Method	Removal, %
Arsenic	Oxidation to As^{5+} required	>90
As^{3+}	Ferric sulfate coagulation, pH 6–8	>90
As^{5+}	Alum coagulation, pH 6–7	>90
	Lime softening, pH 11	>90
Barium	Lime softening, pH 10–11	>80
Cadmium	Ferric sulfate coagulation, pH > 8	>90
	Lime softening, pH > 8.5	>95
Chromium	Ferric sulfate coagulation, pH 6–9	>95
Cr^{3+}	Alum coagulation, pH 7–9	>90
	Lime softening, pH > 10.5	>95
	Ferrous sulfate coagulation, pH 6.5–9	>95
Cr^{6+}	(pH may have to be adjusted after coagulation to allow reduction to Cr^{3+})	
Lead	Ferric sulfate coagulation, pH 6–9	>95
	Alum coagulation, pH 6–9	>95
	Lime softening, pH 7–8.5	>95
Mercury, inorganic	Ferric sulfate coagulation, pH 7–8	>60
selenium, Se^{4+}, Silver	Ferric sulfate coagulation, pH 6–7	70–80
	Ferric sulfate coagulation, pH 7–9	70–80
	Alum coagulation, pH 6–8	70–80
	Lime softening, pH 7–9	70–80

According to type and concentration of the contaminants, precipitation or co-precipitation, or both may be effective in removing metals during chemical coagulation and lime treatment. Some metals will co-precipitate with iron or aluminum hydroxide in coagulation flocculation process. Iron coagulant perform better than the aluminum coagulants in metal removing process since iron hydroxide is insoluble in a wider pH range and is less soluble comparing to the aluminum hydroxide (Pontius, 1990).

KEY POINTS

Coagulants are used in water treatment mainly for destabilization of colloidal suspensions and agglomerate the particles to form settleable flocs; they are dispersed into the treated water by intensive agitation. They are also used to form precipitates that adsorb natural organic matters and some inorganic materials such as phosphates, arsenic compounds and fluoride. Aluminum sulphate (alum), ferric iron salts, organic polymers (polyelectrolytes) are widely used as coagulants in the water treatment plants. Oxidative chemicals such as chlorine, ozone or potassium permanganate can be added prior to the primary coagulant to improve the coagulation by oxidizing dissolved organics. Adjustment of the pH-value may also be considered as a coagulant aid since all the coagulants have an optimum operation pH range.

Flocculation is an operation in which small particles agglomerate into well defined flocs by gentle mixing for a longer time. The flocs can be effectively removed in the following separation processes such as sedimentation, flotation and filtration.

Heavy metals existing as cations in water form insoluble hydroxides and carbonates and precipitate. This is achieved usually by adding caustic soda or lime to adjust the pH according to the maximum metal insolubility.

REFERENCES

Baruth E. E. (1990). Water Treatment Plant Design. 4th edn, AWWA, ASCE. McGraw Hill Handbooks, New York, Chapter 6, 6.1–6.25.

Drinan J. E. (2000). Water and Wastewater Treatment. A Guide for the Nonengineering Professional. CRC press, New York, Chapter 5, 71–77.

Gregory J. (1972). The Scientific Basis of Flocculation, The Netherlands.

Hudson H. E. (1981). Water clarification processes practical design and evaluation, Van Nostrand Reinhold Environmental Engineering Series, 101–121.

Letterman R. D. (1999). Water Quality and Treatment, A Handbook of Community Water Supplies. 5th edn, American Water Works Association, McGraw-Hill, Inc., New York, 6.1–6.61.

Melia C. R. O. (1982). Physicochemical Processes for Water Quality Control. Wiley Interscience, New York.

Metcalf & Eddy Inc. (1990). Wastewater Engineering Treatment, Disposal, Reuse. Revised by George Tchobanoglous, 2nd edn, Tata McGraw-Hill, New York, 76–82.

Niquette P., Monette F., Azzouz A. and Hausler R. (2004). Impacts of substituting aluminum-based coagulants in drinking water treatment. *Water Qual. Res. J. Canada*, **39**(3), 303–310.

Pontius F. W. (1990). Water Quality and Treatment, A Handbook of Community Water Supplies. 4th edn, American Water Works Association, McGraw-Hill, Inc., New York, 269–365.

Tillman G. M. (1996). Water Treatment. Troubleshooting and Problem Solving. Lewis Publishers, Boca Raton, Chapter 5, 49–63.

Chapter 5

Sedimentation and flotation

Ali Tor, Senar Ozcan and Mehmet Emin Aydın

5.1 DESCRIPTION OF SEDIMENTATION

Sedimentation is a physical water treatment process, which is used for removing the suspended solids (i.e. particulate matter, chemical floc, precipitates in suspension and other solids) that are heavier than water (Drinan, 2000). The sedimentation process is based upon the settlement of the suspended solids by gravity and consequenly settleable solids are removed from water as sludge at the bottom of a-circular or a-rectangular tank. An effective sedimentation process removes as much of the floc and other suspended material as possible prior to filtration (Willis, 1990). An example for the diagram of water treatment including sedimentation process in City of Greensboro (North Carolina, USA) is presented in Figure 5.1. (http://www.greensboronc.gov/departments/Water/water-system/diagram.htm). In order to understand the settling in this process, the sedimentation tank can be separated into four zones as briefly described below and as shown in Figure 5.2 (Tillmann, 1996).

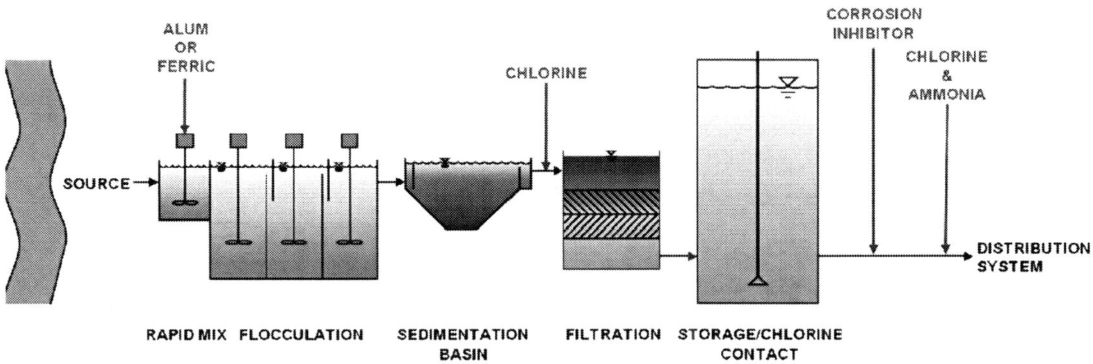

Figure 5.1 An example for the diagram of water treatment including sedimentation process in Greensboro (North Carolina, USA).

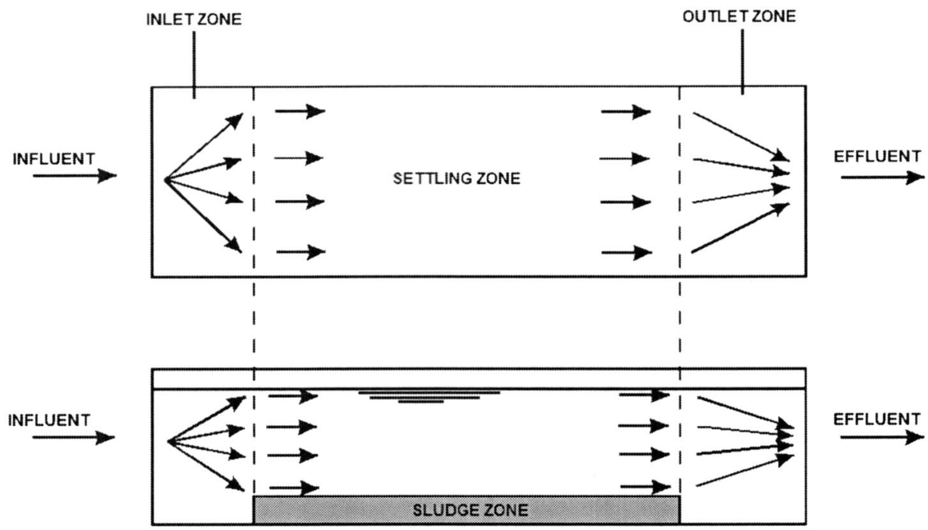

Figure 5.2 Different zones in a typical-rectangular sedimentation tank.

Inlet zone. The inlet or influent zone should provide a smooth transition from the flocculation to the sedimentation tank. It should also distribute the flow uniformly across the the tank. The inlet zone includes baffles which gently spread the flow across the total inlet of the tank and prevent short- circuiting in the tank. Short-circuiting is a term used for description a situation when influent water exits the tank quickly.

Settling zone. The settling zone is the largest section of the sedimentation tank. This zone provides the quiet area required for settlement of the suspended particles.

Sludge zone. The sludge zone is located at the bottom of the tank. This zone provides a storage area for the sludge prior to removal of sludge for additional disposal. The inlet of the sedimentation tank has to be designed to prevent high flow velocities near the bottom of the tank. If the water is allowed to enter the sludge zone with high velocity, the settled sludge may be swept up and out of the tank. Sludge is removed from the sludge zone by scraper which move along the bottom of the tank for further treatment.

Outlet zone. The outlet zone (launder), which has to provide a smooth transition of clarified water from the sedimentation zone, generally consists of weirs. This area of the tank also controls the depth of water in the basin. Weirs placed at the end of the tank are used for both checking the overflow rate and prevention of solids leaving the tank before they settle out. The enough weir length should be employed to control the overflow rate.

5.2 DESIGN APPROACH

General parameters used for the design of sedimentation tank are surface loading rate, retention time, tank depth and flow rate and tank number (Willis, 1990).

Surface loading rate

Surface loading rate or overflow rate is the parameter used to design the sedimentation tank. This rate is defined as the flow rate divided by the tank surface area. Unit is the cubic meter per hour per square meter ($m^3/m^2/h$). Acceptable surface loading rate change with hyraulic characteristics of the sedimentation tank.

Retention time

Retention time can be defined as tank volume divided by the flow rate. In order to obtain an excellent treatment performance, 1.5 to 2 hours should be used as a retention time for conventional sedimentation tanks.

Tank depth and velocities

Although it is expected that depth should not be an important parameter because settling is based on overflow rates, in practice, it is important parameter because it influences flow-through velocity. Flow-through velocities must be low enough to prevent rising of the settled flocs. Velocities of 0.6 to 1.2 meter per minute are usually acceptable for depths of 2.1 to 4.3 m. Furthermore, tank depth might also play a role in allowing greater opportunity for flocculent particle contact. When particles settle down, additional flocculation takes place and this allows growing of heavier floc and formation of a sludge blanket which is less susceptible to resuspension. The formation of this blanket increases the solids content of the residuals withdrawn by removal equipment. Contrary, the blanket can also contribute to the formation of a density current along the bottom of the tank, which causes the floc carryover to the effluent.

Number of tanks

The choice of number of tank is also important parameter when it is design. The minimum and, by far the least costly, plant would have only a single settling tank. But, this would cause poor operation since the tank has to be periodically maintained. While, a minimum of four tanks is prefered. The number of tanks is also depending on the maximum tank size in which the selected sludge removal equipment is accommodated or on other factors, such as area requirement.

Types of sedimentation tank

There are many types of sedimentation tanks. They can be rectangular or circular. Choice of sedimentation tank type is affected by following factors, that is, availability of space, experience and judment of the engineer, size of installation, local site conditions, economics involved, and so on (Eroğlu, 2008).

Rectangular tanks are prefered especially for large-scale water treatment plants. In general, rectangular tanks are designed as being long and narrow, with length-to width ratios of 20 or more. Such a high-value ratio may not be economically acceptable, and a lower ratio as low as about 5 may give acceptable efficiency if the influent is well distributed. This shape allows the least short-circuiting in a tank when actual flow time of water through the tank is less than the calculated time. Short-circuiting is primarily caused by uneven flow distibution and density that form zones of near-stagnant water in corners (Hamlin & Abdul Wahab, 1970). Basin widths are selected by considering the equipment required for sludge removal system. Tanks equipped with mechanical sludge removal systems have deeps usually between 3.0 and 4.3 m (Willis, 1990). The outlet system for a rectangular tank generally is designed using a weir that spills into the effluent flume extending across the entire width of the tank. For chemically treated waters, the outlet system of the tank may also have an orifice providing a suitable flow rate to keep the fine floc.

Circular sedimentation tanks became more widespread in water treatment when periodic manual cleaning of long. This type of tank share some of the performance advantages of the rectangular ones, but are generally more prone to short-circuiting and particle removal problems. Although some circular tanks are designed as being rim feed with clarified water collected in the center, most of them are the center-feed type. Circular tanks have been built as large as 91 m in diameter but typically are less than 30 m in diameter. Typical depths range from 3.0 to 3.6 m. In circular tanks, the flow enters either the center of the tank (center feed) or the side of the tank (side feed). The sludge is usually collected in a hopper near the center of the tank (Willis, 1990; Drinan, 2000; Eroğlu, 2008).

Recently, coagulation, flocculation, and sedimentation have been carried out in a single tank to obtain a compact and relatively economical water clarification. These types of clarifiers are termed as "upflow clarifiers or sludge-blanket clarifiers" because the water flows up toward the effluent launders as the suspended solids settle. In the "sludge-blanket clarifier" or "floc-blanket clarifier", chemically coagulated water is fed upwards through a tank at a rate that flocculation happens and the settling floc particles are held in suspension. At steady state conditions, there is a balance between the upflow rate of the water through the clarifier and the settling velocity of the floc particles. The suspended floc particles form a dense blanket trapping additional particulate matter. The efficiency of the floc blanket clarifier depends on the blanket concentration which is governed by the upflow rate, temperature and floc density (WRc, 2002).

Lamella settler is also commonly used in water treatment. It is an inclined plate and shallow depth sedimentation unit. The most significant aspect of its design is its available settling area. The lamella setler units are design on the basis of square feet of settling area, but using the inclined plates, effective settling area becomes the area of each plate projected on a horizontal surface, multiplied by the number of plates. Using a series of inclined paralelled plates reduce the land area required. When compared to a conventional type settling tank or clarifier, the lamella setler unit uses only 10% of the area, but provide the same efffective total area (Wenk, 1990). A more recent innovation in water clarification involves ballasted flocculation. The trade-named of this process is ACTIFLO®. In this process, the weight of the flocs is increased by causing the attachment of grains of high-density microsand (about 2700 kg/m^3). The efficiency of this process is equal to or better than that of the current conventional treatment and requires less space. The process occurs in four tanks, one for each part of the process: coagulation, injection of microsand and polymer, maturation of flocs and lamella settling (Desjardins *et al.* 2002).

5.3 ADVANTAGES AND DISADVANTAGES OF SEDIMENTATION

The major advantage of sedimentation is that it uses only gravitational force of earth to separate the flocs from water. Moreover, cost effectiveness and low energy consumption are the main advantages of sedimentation process. The major disadvantages of sedimentation are the long separation time required and the large amount of land area required.

Sludge removal

The sludge should be removed from the sedimentation tank due to the following reasons:

- To prevent scouring of solids back into suspension,
- To prevent sludge from becoming septic or imparting a taste or odor to the water,
- To prevent reducing the volume of the basin which reduces the retention time,
- To prevent the accumulation of organic material from possibly trihalomethane formation when chlorine is applied for disinfection (Drinan, 2000).

How frequent sludge removal is required is depend on both amount of solids removed and nature of the floc.

Box of key fact points (Drinan, 2000)

- ***Particle shape.*** A round particle settles easier than a particle having irregular edges.
- ***Time.*** In general, 2 to 4 h is used as retention time for sedimentation.
- ***Flow rate.*** Usually, 0.30 to 0.90 m per minute is used as flow rate.
- ***Character of suspended material.*** A dense, heavy sand or silt particle settles more quicker than a light particle of floc. Furthermore, the efficiency of the coagulation and flocculation process influences the settling characteristic of the floc.
- ***The inlet and outlet arrangement of the tank.*** The inlet parts has to distribute the incoming water over full cross section of the tank. The outlet structure should evenly collect the water from the surface.
- ***Temperature.*** When temperature of water is decreased, flocs generally settle more slowly.

5.4 DESCRIPTION OF FLOTATION

Flotation process is an alternative technique to the sedimentation process. This technique uses gas bubbles to increase the buoyancy of suspended solids. The gas bubbles attach to the particles and make their effective density lower than that of the water. This causes the particles to rise through the water to float to the top. The flotation tank is divided into two zones, contact zone and separation zone (Figure 5.3). The former is the place where particles contact and adhere to bubbles, and the latter zone provides relatively quiescent conditions for particle – bubble agglomerates rising to the surface. Once the particles have reached the surface, they can be collected by a skimmer. However, flotation requires careful control to achieve high quality output. The principal advantages of flotation over sedimentation are that very small or light particles that settle slowly can be removed more completely and in a shorter time. "Float" can be described as a sludge that accumulates on the flotation tanks. The removal of float can be performed by floating and/or mechanical scraping (Rubio *et al.* 2002).

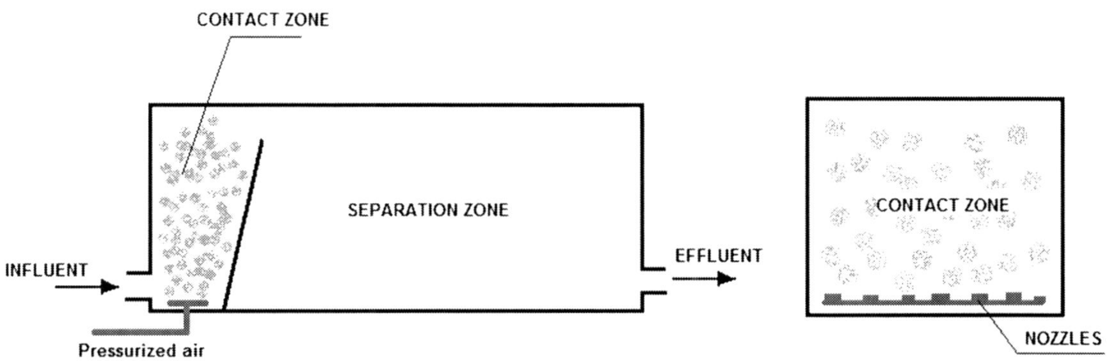

Figure 5.3 General presentation of flotation procedure.

In flotation process, bubbles are formed by a reduction in pressure of water pre-saturated with air at pressures higher than atmospheric. The supersaturated water is forced trough needle-valves or special orifices, and clouds of bubbles, 30–100 µm in diameter, are produced just down-stream of the constriction. Dissolved air flotation (DAF) was recognized as a method of separating particles for many applications including clarification of refinery wastewater, wastewater reclamation, separation of solids and other in drinking water treatment plants and removal/separation of ions. Table 5.1 shows the removal of $Fe(OH)_3$ from water by using batch DAF system (Féris *et al.* 2000). Decreasing the air/water surface tension by addition of a surfactant into the DAF system highly improved the separation process. Féris *et al.* reported that without surfactants, the minimum pressure required for removal of $Fe(OH)_3$ was found to be 3 atm. While, using 30.5 mg/L of sodium oleate reduced the operation pressure to lower than 3 atm.

Table 5.1 Removal of $Fe(OH)_3$ from water by batch DAF system (pH of water: 7, with no surfactant) (Féris *et al.* 2000).

Pressure, atm	Ion concentration, mg/L		Removal efficiency (%)
	Initial	Final	
1.8	30.6	30.6	0
2	29.9	29.9	0
3	29.6	2.1	93.0
4	29.9	1.0	96.5

The types of flotation tanks can be circular (rare for water treatment) or rectangular. Prior to the flotation, preflocculation step is required. In comparison to the circular tanks, the rectangular ones have some advantages that is, scale-up, simple design and easy float removal and relatively small are requirement. Therefore, most of flotation plants prefered the rectangular flotation tanks for water treatment purposes. Rectangular floatation tanks are generally designed with a depth of approximately 5 m and overflow rates of 8–12 m/h. In general, retention time in the flotation tank is between 5–15 minutes (Rubio *et al.* 2002).

5.5 ADVANTAGES AND DISADVANTAGES OF FLOTATION

In comparison to the sedimentation, the major advantage of DAF is that DAF can perform the operation in a smaller space with a smaller piece of equipment. The separated solids are typically more concentrated than "settled" solids. The dissadvantages of DAF systems is that they require power to generate the dissolved air.

KEY POINTS

Metals precipitates or metals adsorbed onto coagulated particles can be removed by both sedimentation and flotation.

Sedimentation is a physical water treatment process, which is used for removing the suspended solids. The sedimentation process is based upon the settlement of the suspended solids by gravity and consequenly settleable solids are removed from water as sludge at the bottom of a-circular or a-rectangular tank. General parameters used for the design of sedimentation tank are surface loading rate, retention time, tank depth and flow rate and tank. The major advantage of sedimentation is that it uses only gravity to separate the flocs from water and is therefore cost effectivene with a low energy usage. The major disadvantages of sedimentation are the long separation time required and the large amount of land area required.

Flotation process is an alternative technique to the sedimentation process. This technique uses gas bubbles to increase the buoyancy of suspended solids. To obtain efficient flotation, the pH and chemical dosage of the coagulation process should be optimized. The time and degree of agitation used for flocculation process also affect the performance of flotation process. Varying the amount of air introduced into the flotation tank affects the flotation efficiency. Therefore, different nozzle sizes require different combinations of flow and pressure to deliver the same amount of air into the flotation system.

REFERENCES

Desjardins C., Koudjonou B. and Desjardins R. (2002). Laboratory study of ballasted flocculation. *Water Res.*, **36**, 744–754.

Drinan J. E. (2000). Water and Wastewater Treatment. A Guide for the Nonengineering Professional, CRC press, Lancaster, PA, Chapter 6, 79–82.

Eroğlu V. (2008). Water Treatment, Ministry of Environment and Forestry (Turkey), Ankara.

Féris L. A., Gallina S. C. W., Rodrigues R. T. and Rubio J. (2000). Optimizing dissolved air flotation design system. *Braz. J. Chem. Eng.*, **17**(4–7), 549–556.

Hamlin M. J. and Abdul Wahab A. H. (1970). Settling characteristics of sewage in density currents. *Water Res.*, **4**, 609–610.

Otter (2002). Process Model Descriptions, Floc Blanket Clarifier. 19–26, WRc plc, Great Britain.

Rubio J., Souza M. L. and Smith R. W. (2002). Overview of flotation as a wastewater treatment technique. *Minerals Eng.*, **15**, 139–155. (http://www.greensboro-nc.gov/departments/Water/watersystem/diagram.htm) (2009).

Tillman G. M. (1996). Water Treatment. Troubleshooting and Problem Solving. Lewis Publisher, Boca Raton, Chapter 6, 69–72.

Wenk S. E. (1990). The theory, design and experience of Lamella gravity settlers in the phosphate industry. *Nutr. Cycling Agroecosyst.*, **25**, 139–143.

Willis J. F. (1990). *Water Treatment Plant Design.* 4th edn, AWWA, ASCE. McGraw-Hill Hanbooks, New York, Chapter 7, 1–43.

Chapter 6

Removal of metals from drinking water by filtration

Lary Russell and Todd Russell

6.1 INTRODUCTION

Filtration is an important component of metal removal in many small water systems. Small water systems frequently combine oxidation of metals with removal by sedimentation and filtration. A variety of metals in drinking water will respond to oxidation by precipitation and removal, such as, iron, manganese, chromium III, selenium, arsenic III. However, by far the most common metals of concern in drinking water are iron and manganese. The reduced forms of these metals [Fe(II) and Mn(II)] present similar challenges when found in solution and often appear in tandem. However, the differing chemistries and reaction kinetics between iron and manganese has impacts on filter design. Removal of iron and manganese by filtration for small water systems is the primary subject of this chapter.

As noted in Chapter 3, the United States Environmental Protection Agency (USEPA) set the secondary maximum contaminant levels, SMCLs, for iron and manganese at 0.3 mg/L and 0.05 mg/L and the EU sets the standards at 0.2 and 0.05mg/L, respectively (USEPA, 1992). The American Water Workers Association (AWWA) suggests even more stringent limits of 0.05 mg/L of iron and 0.01 mg/L of manganese. In many small water systems with iron and manganese in their source water, filtration is a key component to meeting regulatory standards.

6.2 FILTRATION OVERVIEW

Chemical oxidation is one process that can be used to remove dissolved iron and manganese from drinking water. The premise of chemical oxidation is to convert dissolved Fe(II) and Mn(II) to insoluble ferric hydroxide and magnesium oxides. The precipitate can then be removed from solution by filtration or sedimentation followed by filtration. A typical treatment train consists of an oxidant dose followed by a detention basin and finally rapid filtration or pressure filtration.

For systems utilizing oxidation for removal of iron and manganese, proper selection of the filter media and design of the filtration unit is important. Filters play a critical role in most of these systems as they can both remove flocculated metal oxides and can catalyze further metal removal. Many times the low concentrations of metals in source waters make the formation of settleable metal oxide flocs challenging.

However, it is much more feasible to form metal oxides that are filterable. This process can be visualized where water with dissolved iron and or manganese will appear clear after aeration, but still contains microscopic colloidal iron and manganese. Only once reaching the filter will these colloidal oxidized metals precipitate out onto the media forming a dark red or black color (MWH, 2005). Filter media commonly used for removal of iron and manganese include coal, sand or manganese greensand. The following sections outline the use of these different media for metal removal.

6.3 THE AUTOCATALYTIC REACTION OF MANGANESE

One of the major differences in the removal of iron and manganese is the autocatalytic reaction of manganese. This autocatalytic reaction takes place in the filter and is critical for manganese removal. The reaction kinetics and chemistry of manganese make the formation of a settleable manganese oxide flocs challenging on timescales that are practical for small water treatment plants. Instead manganese is removed primarily in the filter by the autocatalytic reaction with previously deposited manganese oxides. The autocatalytic nature of manganese oxides on filter media adsorbs Mn(II) from solution. The manganese oxides on the media then can catalyze the reaction to form manganese oxides from adsorbed Mn(II). Through this process manganese is removed by the combination of adsorption of dissolved Mn(II) and the formation of oxides. Due to these combined process manganese in filters is typically found in non-stoichiometric ratios ranging in degrees of oxidation from $MnO_{1.3}$ to $MnO_{1.9}$ (MWH, 2005). To take advantage of the autocatalytic nature of manganese proper filter media selection and filter break-in are important. For smaller systems designed to remove manganese it is common to use media such as greensand which is preconditioned to have a permanganate or manganese oxide coating. A further discussion of this media is presented in the ‘Greensand’ section below.

6.4 FILTER HYDRAULICS AND BACKWASHING

The filter media choices play an integral role in the hydraulics behavior of the filter. In gravity driven systems, the limited head is an important design constraint on the rapid filter media. While the filter plays an integral role in the precipitation and removal, it is rare to observe excessive build up of oxides on the filter media. However, in waters containing high concentrations of iron and manganese, suspended solid accumulation in the filter can be a concern (JMM, 1985). In waters containing high concentrations of iron sedimentation prior to filtration may be necessary. Filtration media typically have effective sizes >1.5 mm and low uniformity coefficients. Typical loading rates for these filters range from 300–600 m/d (5–10 gpm/ft^2) and backwashing rates range from 900–1500 m/d (15–25 gpm/ft^2) (JMM, 1985). The ability to efficiently backwash the media is a valuable characteristic for filters. The media used will impact both the backwashing frequency and the required backwashing velocities. Medias that require less backwashing velocity are advantageous as they tend to produce less backwash water. Depending on how the backwashing water is handled, lower backwashing velocities can be an important design consideration. Approximately three quarters of the backwash water can recovered after the precipitated iron and manganese are settled (JMM, 1985).

Multimedia filters can also be considered depending on the removal demands and hydraulics of the system. In rapid filtration systems multiple layers can be used to allow for removal of a wider variety of particulate sizes. In treatment plants which are utilizing filters for both metals removal and other treatment process, such as, flocculation with chemical coagulation, a multilayer filter can be an attractive choice to avoid clogging while still removing the metals of concern. In systems designed to primarily remove iron and manganese, the wider range of particle capture may not be an advantage.

In systems utilizing potassium permanganate, as on oxidant, it is common to utilize pressure filtration. These filters typically run at loading rates ranging 250–500 m/d (4–8 gpm/ft^2) and backwashing rates ranging from 500–1200 m/d (8–20 gpm/ft^2) (MWH, 2005). Pressure filters used for metal removal are typically run with greensand, but can contain other media. Greensands are natural zeolites and they tend to have significantly smaller effective sizes, which generate higher head losses. However, the chemical properties of the greensand can provide advantages that warrant its use, particularly in systems using potassium permanganate as an oxidant.

6.5 COAL AND SAND

Coal and sand are both common media choices for rapid filtration used in the removal of iron and manganese. For metal removal these media should typically have effective sizes >1.5 mm and low uniformity coefficients (JMM, 1985). Typically coal is good media choice for this application due the ability to backwash with lower velocities (JMM, 1985) and lower headloss for a given filter bed height. The decision of the best media to use may rely on the other treatment process which the filter is involved in besides metal removal. When implemented correctly, coal or sand filters will acclimate within three weeks and produce high quality effluent with low iron and manganese (JMM, 1985).

Rapid sand filters utilized for iron removal can have issues with the growth of bacterial slimes within the filter. A pilot study using aeration, reaction-sedimentation basin and rapid sand filtration found that heavy bacterial growth promoted anaerobic conditions within the filter (Faust & Aly, 1998). If anaerobic conditions exist in the filter, previously oxidized metals can be reduced allowing them back into solution. Rapid filters should be monitored for the bacterial growth which could counteract metal removal.

6.6 GREENSANDS

Greensand is a naturally forming zeolite which has chemical properties that causes more rapid oxidation and removal of iron and manganese. Greensand has a smaller effective size (<0.3 mm) (MWH, 2005) than sand and coal which causes it have much higher head losses when used as a filter media. These higher head losses mean that greensand is run in a pressure filter rather than a gravity filter. Pressure filtration can be expensive and requires an additional capital investment. Because of these costs the use of greensand is generally not justified in larger water treatment plants. However, the chemical properties of greensand make it appealing for small water treatment plants (MWH, 2005). Filter depth in greensand filters is similar to conventional filters.

One of advantages of greensand lies in the ability to precharge it with permanganate on an intermittent basis. The periodic recharging can be done with a $KMnO_4$ (potassium permanganate) solution. During recharging, the greensand media adsorbs the permanganate. The adsorbed permanganate can then act to further adsorb iron and manganese out of solution. In small systems, it is possible to periodically recharge the greensand filter media and achieve metal removal without a continuous chemical feed (JMM, 1985).

If utilizing a continuous feed of $KMnO_4$ as the oxidant for metal removal, it is important to consider the use of greensand due to its ability to remove $KMnO_4$ from solution (JMM, 1985). Excess $KMnO_4$ in solution (>0.05 mg/L) (MWH, 2005) turns the water pink and is easily observed by consumers. To avoid consumer complaints and remove iron and manganese, requires correct dosing of $KMnO_4$. The ability of greensand to absorb $KMnO_4$ allows for more flexibility in chemical dose without producing excess $KMnO_4$. Correct dosing can be addressed in a pilot study, which accounts for possible seasonal variations in iron and manganese concentrations in the source water.

Greensand has also been commonly use in treatment systems for iron and manganese removal that utilize a combination of Cl_2 and $KMnO_4$ as oxidants. These systems use prechlorination, alum coagulation, sedimentation and filtration. In these systems the filter media is conditioned with $KMnO_4$ to facilitate the formation of a manganese oxide on the surface. The manganese oxides on the surface of the media adsorb Mn(II) and Fe(II). A dose of chlorine just prior to the filter can be adjusted to facilitate the formation of oxides and enhance removal in the conditioned filter (MWH, 2005).

6.7 PILOT TESTING

The removal of metals by oxidation followed by filtration is a complex process whose characteristics are hard to predict. A wide variety of source water characteristics can impact both the oxidation process and filtration effectiveness. While the principles of iron and manganese removal are well established, correctly predicting the behavior of a system with a given source water is challenging. When feasible, to address these design challenges pilot scale testing is strongly encouraged. For filtration, the pilot test can provide valuable insights into the best media choices, the filtration rates and the filter backwashing rates and frequency (JMM, 1985). An investment in filter pilot testing can provide significant returns in proper design, enhanced removal and lower operating costs.

KEY POINTS

Filtration is an important component of metal removal, frequently combining oxidation of metals with removal by sedimentation and filtration. The most common metals of concern in drinking water are iron and manganese. Although the reduced forms of these metals [Fe(II) and Mn(II)] present similar challenges when found in solution, the differing chemistries and reaction kinetics between iron and manganese has impacts on filter design.

Chemical oxidation converts dissolved Fe(II) and Mn(II) to insoluble ferric hydroxide and magnesium oxides. The precipitates can then be removed from solution by filtration or sedimentation followed by filtration. A typical treatment train consists of an oxidant dose followed by a detention basin and finally rapid filtration or pressure filtration.

Filters can both remove flocculated metal oxides and can catalyze further metal removal. One of the major differences in the removal of iron and manganese is the autocatalytic reaction of manganese. This autocatalytic reaction takes place in the filter and is critical for manganese removal.

For systems utilizing oxidation for removal of iron and manganese, proper selection of the filter media and design of the filtration unit is important. Filter media (type and size), filter area and depth, hydraulic and solids loading rate and backwashing regimes are all important aspects of filter design. An investment in filter pilot testing can provide significant returns in proper design, enhanced removal and lower operating costs.

REFERENCES

AWWA (2003). Water Quality: Principles and Practices of Water Supply Operations. American Water Works Association, Denver.

Faust S. D. and Aly E. M. (1998). Chemistry of Water Treatment. 2nd edn, Ann Arbor Press Inc., Chelsea, Michigan.

HDR Engineering, Inc. (2001). Handbook of Public Water Systems. 2nd edn, John Wiley & Sons, Inc., Hoboken, New Jersey.

J. M. Montgomery, Consulting Engineers Inc. (1985). Water Treatment Principles and Design. John Wiley & Sons, Inc., New York.

MWH (2005). Water Treatment: Principles and Design. 2nd edn, John Wiley & Sons, Inc., Hoboken, New Jersey.

USEPA (1992). Secondary Drinking Water Regulations: Guidance for Nuisance Chemicals. United States Environmental Protection Agency, Washington, D.C. [http://www.epa.gov/ogwdw000/consumer/2ndstandards.html] [accessed June 2010].

Chapter 7

Electrochemical treatment methods

Ona Gylienė

7.1 THEORETICAL BACKGROUND OF THE ELECTROCHEMICAL PROCESSES

Electrochemical treatment is an emerging technology used for the removal of organic and inorganic impurities from water and wastewater.

Electrochemical processes involve redox reactions, where oxidation and reduction reactions are separated in space or time. Usually the electrochemical treatment of water is concerned with electron transfer at the solution/electrode interface applying an external direct current in order to drive an electrochemical process. The electrochemical reactions proceed in an electrochemical cell using electrodes – cathode and anode. The cathode and anode are defined by convention, with the anode always being the site of oxidation and the cathode the site of reduction. The solid electrodes have been applied mainly in treatment processes. Two types of current may flow in an electrochemical cell, faradaic and non-faradaic. Faradaic currents are those which are created by the reduction and/or oxidation of the chemical species in the cell. Non-faradaic currents are related to the resistance of cell.

All electrochemical reactions occur in the medium containing dissolved ions (electrolyte), which are mobile and able to support current flow and to diminish the unproductive non-faradaic current. The most popular solvent for electrolytes used in electrochemistry is water and the electrolytes – sodium sulfate and chloride.

Electrochemical methods have a long history as water treatment technologies for removal of a wide range of pollutants. However these methods have never become accepted as a "mainstream" of water treatment technology. Nowadays technical improvements combined with a growing need for small-scale decentralised water treatment facilities have led to a re-evaluation of the electrochemical treatment methods.

7.2 ELECTROLYSIS

Electrolysis is a process in which one species in solution (usually a metal ion) is reduced by electrons at the cathode and another gives up electrons to the anode and is oxidized there. When the electrolysis is applied to treatment of drinking water, where the cation and anion concentrations are low, the main reaction taking place at the cathode in water solutions is hydrogen evolution.

$$2H_2O + 2e^- \longrightarrow H_{2(g)} + 2OH^-$$

while at the anode the evolution of oxygen proceeds

$$H_2O \longrightarrow 2H^+ + \tfrac{1}{2}O_{2(g)} + 2e^-$$

When the water contains dissolved chloride ions, the evolution of chlorine onto the anode also takes place

$$2Cl^- \longrightarrow Cl_{2(g)} + 2e^-$$

The electrodes used must possess a high electrochemical conductivity and be chemically and electrochemically stable. A number of metals and metal alloys can be used as a cathode; meanwhile the number of conductive materials, which are stable to anodic polarization and the action of Cl_2 and O_2, is very limited. Usually, carbon felt, carbon fibers, diamond, platinum, titanium coated with IrO_2/RuO_2 are used as anodes. These anodic materials can generate hydroxyl radicals and other oxidizing species, such as ozone and hydrogen peroxide, which can mineralize the organic pollutants or convert the toxic substances to biodegradable compounds. The identification of chlorine as a source of potentially harmful disinfection by products, and the emergence of recalcitrant pathogens has led to heightened regulation for the removal of microbial pathogens and disinfection by-products from drinking water. As a result, research and development of alternative disinfection technologies has intensified. Electrochemical disinfection has emerged as one of the more feasible alternatives to chlorination (Pillai *et al.* 2009; Bergmanan & Koparal, 2005). Usually in practice large scale electrolysis is applied for generation of strong oxidation agents, which are used for disinfection of drinking water instead of much more hazardous chlorine.

In municipal drinking water supply facilities the direct treatment of drinking water by means of electrolysis is scarcely used. However, electrolysis for preparing drinking water in relatively small quantities (for caravans, camps, remote rural areas etc.) and domestically is widely used in different countries (Germany, USA, Russia, India etc.). For this purpose the different types of cells, electrodes, treatment conditions and so on are arranged in different types of filters (Karnik *et al.* 2005; Appleyard, 2009). They are produced and used world-wide to improve the quality of drinking water. In the year of 1990 household EMERALD (IZUMRUD, Russia) devices destined for water purification and conditioning were put by Laboratory of Electrotechnology, ltd. in serial production. EMERALD devices are unique and have no analogues. Electrochemical reactors constructed of patented MB-11 diaphragm flow electrochemical modular elements are the main elements of these devices.

As evolution of H_2 and Cl_2 gases are the main reactions in electrolysis, the water during treatment becomes alkaline. Therefore this water is usually denominated as "ionized water" or "alkaline water".

Example of the electrochemical treatment of tap water using one camber cell and platinum electrodes at current density 1 A/cm^2 is presented in Figure 7.1. With increase in treatment time the hardness and the content of heavy metals in water decreases, meanwhile the pH (alkalinity) of water increases.

The producers of electrolyzers glamorize the health-heaving properties of such treated water. According to the producers, electrochemically treated water is able to cure a number of dangerous illnesses; the taking of such water allows everyone to feel better, younger, more vigorous, and so on. The reasons of health effects of electrochemically treated water could be very different. If the potable water is free of mineral impurities, no significant effects of electrolysis will occur, because of a very small number of dissociated ions in pure water. Pure water conducts electric current very poorly, so the process is extremely slow and inefficient. If the water contains dissolved salts or "hardness ions" such as calcium or magnesium in medium concentrations, electrolysis might be possible with effective removal of certain impurities. In case of applying high current densities such as in catalytic conversion water filters, which use additional amounts of electrolytes, practically all hazardous substances such as chlorine, organic matter, heavy metals and microorganisms are converted into harmless oxidized forms (Khanniche *et al.* 2001).

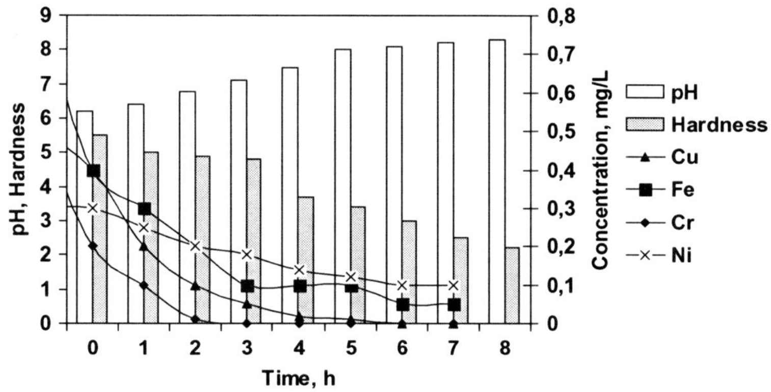

Figure 7.1 Electrolysis of tap water.

The electrolyzing equipment also helps sanitize the drinking water. Between the electrodes naturally present substances in water are converted into oxidizing or disinfecting components. In water containing minerals and chloride ions (20 mg/l), the electrolysis creates sodium hypochlorite. Sodium hypochlorite kills a variety of germs and is widely used to disinfect drinking water. A further advantage is that the germs cannot become resistant to the sodium hypochlorite.

The electrochemical treatment technology presents different *advantages* such as the versatility of the process and the simplicity of the reactors in terms of construction and management, which makes them particularly suitable for automation. The electrochemical processes are highly intensive. The small size of the electrochemical units allows realizing a mobile fresh water treatment unit in order to use it for preparing of drinking water.

The main *disadvantage* of electrolysis is the necessity of the presence of minerals in quantities corresponding to the quantity of minerals (calcium, magnesium) in drinking water. Non-conducting water cannot be treated electrochemically. Salt water cannot be transformed into drinking water by means of electrolysis, as sodium, potassium, calcium and magnesium are not deposited onto the cathode.

7.3 ELECTRODIALYSIS

Electrodialysis is used to remove substances possessing charge from solution through ion-exchange membranes under the influence of the applied electric potential difference. This is done in electrodialysis cells which are presented in Figure 7.2. The cell consists of a feed compartment and a concentrate (brine) compartment, which is constructed from both the anion exchange and cation exchange membranes. These membranes are placed between two electrodes. In almost all practical electrodialysis processes, multiple electrodialysis cells are arranged into a configuration called an electrodialysis stack, with alternating anion and cation exchange membranes. During electrodialysis substances dissolved in water are removed from the solution. This process is reverse to reverse osmosis, where pure water penetrates through the membrane.

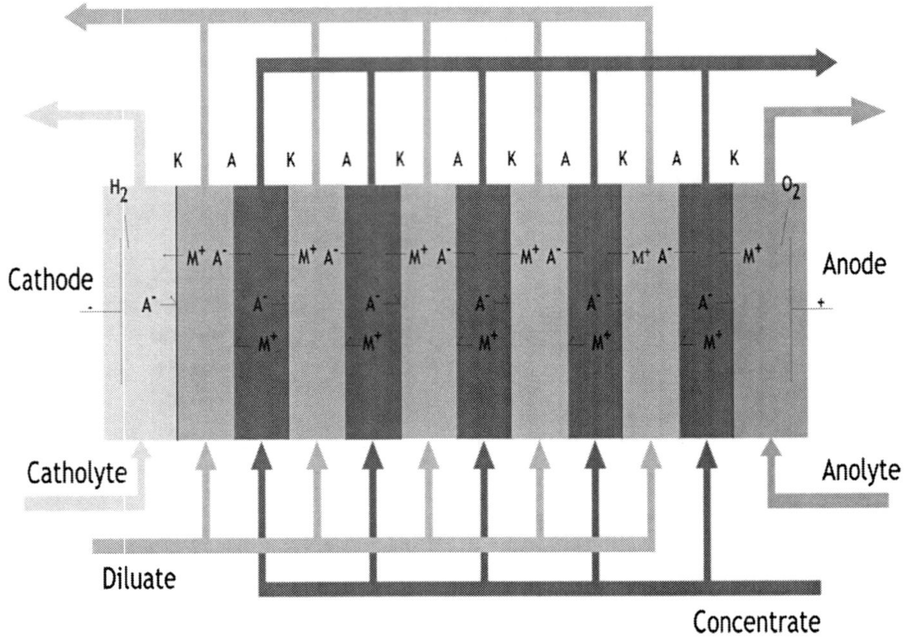

Figure 7.2 Principal scheme of electrodialyser.

Small amounts of hydrogen gas are generated at the cathode and small amounts of either oxygen or chlorine gas (depending on the composition of the solution treated and ion exchange membrane arrangement) at the anode. Usually, the electrodes and membranes are arranged in such a manner, that pH of solution is kept close to neutral.

The major application of electrodialysis has historically been the desalination of brackish water or seawater as an alternative to reverse osmosis for potable water production and seawater concentration for salt production (primarily in Japan) (Shaposhnik *et al.* 2002). Differently from reverse osmosis, when the purified drinking water requires addition of minerals, the electrodialysis enables to prepare drinking water with a required quantity of minerals.

Electrodialysis has inherent limitations, working best at removing the low molecular weight ionic components from a feed stream. Non-charged, higher molecular weight and less mobile ionic species will not be significantly removed. Also, in contrast to reverse osmosis, electrodialysis becomes less economical when extremely low salt concentrations are present in water. Current density becomes limited and current utilization efficiency typically decreases with decrease in salt concentration. Comparatively large membrane areas are required to satisfy capacity requirements for low concentration (and sparingly conductive) feed solutions. Innovative systems overcoming the inherent limitations of electrodialysis (and reverse osmosis) are available; these integrated systems work synergistically, with each sub-system operating in its optimal range, providing the least overall operating and capital costs for a particular application.

When natural waters are treated using electrodialysis, calcium hydrocarbonate removal is restricted to the diffusion current density. Exceeding this value causes irreversible water dissociation at the interface of the anion-exchange membrane and solution and the selective transfer of the hydroxyl ions through the anion-exchange membrane into concentrating cells. These hydroxyl ions accumulated in the diffusion boundary layer at the anion-exchange membrane may interact with magnesium, calcium and hydrocarbonate ions to form magnesium hydroxide and calcium carbonate, thus causing the membrane blocking (fouling). It could be partially avoided by applaying electrodyalysis reversal, when the direction of ion flow is periodically reversed by reversing the polarity of the applied electric current.

In normal potable water production without the requirement of high recoveries, reverse osmosis is generally believed to be more cost-effective when total dissolved solids are 3 g/L or greater, while electrodialysis is more cost-effective for feed concentrations less than 3 g/L or when high recoveries of the feed are required.

Usually, electrodialysis is applied for small and medium scale drinking water production (villages, military camps, nitrate reduction, hotels and hospitals).

The water produced using electrodialysis must also be treated for organic compounds (if they are a concern). Because source water does not physically pass through membranes in these systems, most organic contaminants are not removed. Microbes also remain in water. Therefore microbes must be removed either before or after the electrodialysis process.

The membrane fouling is also a weakness of this treatment process.

7.4 ELECTROCOAGULATION

Electrocoagulation is the electrochemical production of destabilisation agents (usually Al or, Fe ions) that neutralize the electric charge of suspended pollutant. Electrochemically generated metallic ions from these electrodes can undergo hydrolysis near the anode to produce a series of activated intermediates that are able to destabilize finely dispersed particles present in the water/wastewater to be treated. The destabilized particles then aggregate to form flocs (Bennajah *et al.* 2009); Ghosh *et al.* 2008).

The coagulant (Fe or Al ions) is produced, when the electrode is charged positively. Due to the high electronegative potential of Al^{3+} and high overvoltage of Fe^{2+} and Fe^{3+}, the Al or Fe are not deposited onto cathode.

When aluminum is used as an electrode, the reactions are as follows:

At the cathode,

$$2H_2O + 2e^- \longrightarrow H_{2(g)} + 2OH^-$$

At the anode,

$$Al_{(s)} \longrightarrow Al^{3+} + 3e^-$$

In the solution

$$Al^{3+}{}_{(aq)} + 3H_2O \longrightarrow Al(OH)_3 + 3H^+{}_{(aq)}$$

When iron is used as an electrode, the reactions are as follows:

At the cathode,

$$2H_2O + 2e^- \longrightarrow H_{2(g)} + 2OH^-$$

At the anode,

$$4Fe_{(s)} \longrightarrow 4Fe^{2+}{}_{(aq)} + 8e^-.$$

When the oxygen is dissolved in solution, the ferrous hydroxides are formed

$$4Fe^{2+}{}_{(aq)} + 10H_2O + O_{2(g)} \longrightarrow 4Fe(OH)_3 + 8H^+{}_{(aq)}.$$

The overall reaction

$$4Fe_{(s)} + 10H_2O + O_{2(g)} \longrightarrow Fe(OH)_{3(s)} + 4H_{2(g)}.$$

During electrocoagulation an intense evolution of hydrogen gas proceeds. The gas evolved causes the flotation of the flocks. Thus, the electrocoagulation in many cases is undivided from electroflotation. To be precise, this treatment process should be named as electrocoagulation/electroflotation.

The treatment method has proven to be very effective in removing contaminants from water and is characterized by reduced sludge production, no requirement for chemical use, and ease of operation. In the literature the application of electrocoagulation is reported to induce various benefits in comparison to conventional treatments, including environmental compatibility, versatility, energy efficiency, safety, selectivity, ability to automation and cost effectiveness. However, a key limitation is the loss of electroactivity of anode (passivation) in certain pollutants containing solutions. In order to keep the anode active the alternating current is applied. Nowadays the electrocoagulation also is applied for production of electrocoagulants, which are further used as chemical coagulants.

Other reason of limited use of electrocoagulation in practice is that the efficiency of electrocoagulation is strongly dependent on the design and geometry of electrochemical reactors. In its simplest form, an electrocoagulation reactor may be made up of an electrolytic cell with a soluble anode and cathode. The conductive metal plates may be made of the same or different materials (anode and cathode). The current applied can be direct or alternating. The alternating current is usually applied when the both electrdodes are soluble. Alternating current enables to avoid the passivation of electrode.

Coagulation using chemical coagulants is one of the most essential processes in the conventional treatment of drinking water. However chemical coagulation has some inherent problems in cost, maintenance, and sludge production. Electrocoagulation has recently been suggested as an alternative to conventional coagulation. Electrocoagulation can be applied to a broad range of pollutant treatment. It is most effective in removing inorganic (arsenite, arsenate, heavy metals, fluoride, phosphate etc.) contaminants and pathogens.

Electrocoagulation offers significant cost advantages, simplicity of equipment for production of potable drinking water as opposed to other treatment methods such as reverse osmosis and distillation used for desalination of sea water. However, the electrocoagulation is not used in large scale water supply facilities. It is used in decentralized drinking water preparing systems. Usually the application of the electrocoagulation alone does not ensure the high quality of drinking water. It is used as a integrate part of the total treatment system (Holt *et al.* 2005; Emamjomeh & Sivakumar 2009; Zhao *et al.* 2009).

In great numbers such autonomous systems are produced in Sewerodwinsk (Russia) and widely used in practice. The system is composed from the ozonation, filtration, electrocoagulation, UV treatment modules can be completed in dependence of pollutants present in water. The capacity of the systems reaches 50 m^3 a day.

The need for such systems is huge in remote, rural areas, camps, and so on. Such systems are also used in MidEast countries, India, and so on.

In Silchar, Assam (India) electrocoagulation has been used for the treatment of turbid water for drinking purpose for a long time. The first plant of 1350 L/h capacity was given a trial run for more than three months under field condition in 1985. Since then a few trial runs on different quality of water like water containing iron under different field conditions have been carried out with full success.

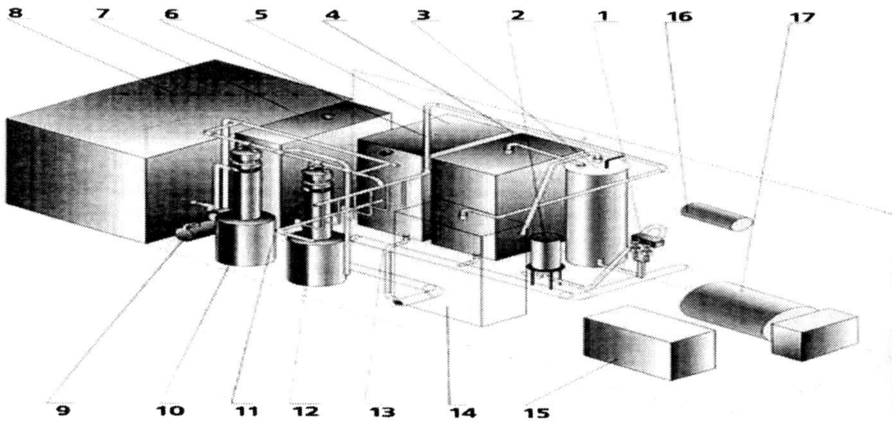

1. Hydrocyclone. 2. *Prefiltration.* 3. *Primary ozonation.* 4. *Electrocoagulation.* 5. *Filtration for floatating load.* 6. *Pipe for ozone.* 7. *Sekondary ozonation.* 8. *Collector.* 9. *Pump.* 10 *Desalination.* 11. *Pipe.* 12. *Microfiltration.* 13. *UV treatment.* 14. *Ozone producer.* 15. *Control desk.* 16. *Heater.* 17. *Diesel dynamo*

The produced electrogoagulant could be used separately for drinking water and wastewater treatment. In lithuanian company INECO produces the electrocoagulant "Hydrogent-SM" for drinking water and wastewater treatment. The results of effectiveness of treatment are presented in Table 7.1.

Table 7.1 Results of potable water and wastewater treatment with "Hydrogent-SM".

Solution	Concentrations, mg/l	
	Before treatment	**After treatment with Hydrogent-SM**
Potable water	Zn – 5,1;Cr – 0.25; Cu – 2,9;Fe – 3.2	Zn < 0,005;Cr < 0,01; Cu < 0,01;Fe < 0.05
Waste water	Zn – 23,1;Cr – 96,0; Cu – 46,0	Zn < 0,005;Cr < 0,01; Cu < 0,01

The structure of electrochemically produced electrocoagulant essentially differs from that of iron or aluminium salts. It is described as polymeric compound with formula $[Fe_2(OH)_n(SO_4)_{(6-n)/2}]_m$, where m depends on n ($n < 2$). It is worth to note that the presence of organic and non-ionic pollutants has insignificant influence on treatment process (Lakshmanan *et al.* 2009; Vasudevan *et al.* 2008; Nikolaev *et al.* 1982). The main part of them is removed with electrocoagulant.

The summary of principal points of electrochemical treatment methods is presented in Table 7.2.

Table 7.2 The main characteristics of the electrochemical methods for drinking water treatment.

Parameters	Electrolysis	Electrodialysis	Electrocoagulation
Equipment	Simple	Complicated	Simple
Contaminant removal	Ionic overwhelmingly	Ionic only	Ionic and nonionic

(Continued)

Table 7.2 The main characteristics of the electrochemical methods for drinking water treatment (*Continued*).

Parameters	Electrolysis	Electrodialysis	Electrocoagulation
Comparable costs	High	Very high	High
Field of application	Household and remote areas	Water treatment plants	Heavy polluted water sources
Outlook for future applications	Increased	Decreased	Increased
Quantity of sludge	Small	Very small	High
Pathogen removal	Complete	Insignificant	Complete

KEY POINTS

The electrochemical treatment methods (electrolysis and electrocoagulation) are not applied in centralized drinking water supply systems. These treatment technologies are most suitable for decentralized water treatment and supply drinking water for small communities in remote areas. Taking into account that the pollution of natural sources of drinking water steadily increases, the electrochemical treatment methods have a future as an advanced technologies for additional treatment of potable water domestically and remote areas.

REFERENCES

Appleyard S. (2009). Assessing the use of simple dye-sensitized solar cells for drinking waterchlorination by communities with limited resources. *Renew. Energy*, **34**, 1651–1654.

Bennajah M., Gourich B., Essadki A. H., Vial Ch. and Delmas H. (2009). Defluoridation of Morocco drinking water by electrocoagulation/electroflotation in an electrochemical external-loop airlift reactor. *Chem. Eng. J.*, **148**, 122–131.

Bergmanan H. and Koparal S. (2005). The formation of chlorine dioxide in the electrochemical treatment of drinking water for disinfection. *Electrochimica Acta*, **50**, 5218–5228.

Emamjomeh M. M. and Sivakumar M. (2009). Review of pollutants removed by electrocoagulation and electrocoagulation/flotation processes. *J. Environ. Manage.*, **90**, 1663–1679.

Ghosh D., Medhi C. R. and Purkait M. K. (2008). Treatment of fluoride containing drinking water by electrocoagulation using monopolar and bipolar electrode connections. *Chemosphere*, **73**, 1393–1400.

Holt P. K., Barton G. W. and Mitchell C. A. (2005). The future for electrocoagulation as a localized water treatment technology. *Chemosphere*, **59**, 355–367.

Karnik B. S., Davies S. H., Baumann M. J. and Masten S. J. (2005). Fabrication of catalytic membranes for the treatment of drinking water using combined ozonation and ultrafiltration. *Environ. Sci. Technol.*, **39**(19), 7656–7661.

Khanniche M. S., Morgan P. G., Khanniche K. N., Jobling C. P. and Khannich N. (2001). A novel electro-chemical process for water treatment. *Rev. Energy Ren.: Power Eng.*, **35**, 63–67.

Lakshmanan D., Clifford D. A. and Gautam S. (2009). Ferrous and ferric ion generation during iron electrocoagulation. *Environ. Sci. Technol.*, **43**(10), 3853–3859.

Nikolaev N. V., Kozlovskii A. S. and Utkin I. I. (1982). Treating natural waters in small water systems by filtration with electrocoagulation. *Sov. J. Water Chem. Technol.*, **4**(3), 244–247.

Pillai K. Ch., Kwon T. O., Park B. B. and Moon Sh. (2009). Studies on process parameters for chlorine dioxide production using IrO_2 anode in an un-divided electrochemical cell. *J. Hazard. Mater.*, **164**, 812–819.

Shaposhnik V. A., Zubets N. N., Strygina I. P. and Mill B. E. (2002). High demineralization of drinking water by electrodialysis without scaling on the membranes. *Desalination*, **145**, 329–332.

Vasudevan S., Sozhan G., Ravichandran S., Jayaraj J., Lakshmi J. and Sheela S. M. (2008). Studies on the removal of phosphate from drinking water by electrocoagulation process. *Ind. Eng. Chem. Res.*, **47**, 2018–2023.

Zhao H. Z., Yang W., Zhu J. and Jin-Ren N. (2009). Defluoridation of drinking water by combined electrocoagulation: Effects of the molar ratio of alkalinity and fluoride to Al(III). *Chemosphere*, **74**, 1391–1395.

Chapter 8

Adsorption processes

Magdalena Zabochnicka-Swiatek, Ona Gyliene, Karin Cederkvist and Peter E. Holm

8.1 INTRODUCTION

Each metal has a tendency to form specific compounds in water and they can be removed by different mechanisms – adsorption or/and ion exchange. The term sorption encompasses both adsorption and absorption, while desorption is the reverse of adsorption. It is a surface phenomenon. Adsorption is the adhesion of atoms, ions, or molecules of gas, liquid, or dissolved solids to a surface forming a molecular or atomic film. The adsorbing species is called the **adsorbate**, and the solid media, to which the sorbate is attracted, is known as the **adsorbent**. Absorption is a process that the atoms, molecules, or ions enter some bulk phase – gas, liquid or solid material.

Adsorption is usually described through isotherms: Freundlich, Langmuir and BET. Langmuir and Freundlich adsorption equations can be used for surfaces which are covered by only one layer of adsorbate and the BET surface area is calculated using the multilayer model.

8.2 FACTORS INFLUENCING SORPTION CAPACITY

The main factors influencing sorption processes are:

(1) Adsorbent characteristics.
(2) Adsorbate properties.
(3) Reaction environment.

In water treatment adsorbents could be used in the form of spherical pellets, rods, moldings, or monoliths. The specific composition and structure of adsorbent influences its sorption capacity. The most important characteristics affecting the sorption capacity of an adsorbent are:

- high surface area,
- high surface charge (positive or negative),
- high thermal and chemical stability.

The sorption capacity is directly proportional to the surface area – the adsorption increases with the increase of specific surface area. The surface charge determines the ability to prefer cations or anions. The positive surface charge provides sites for sorption of anions and the negative surface charge provides sites for cations. The high thermal and chemical stability enables repetitive use at high temperature.

The understanding the adsorbate properties will enable to develop accurate and stable adsorbent–adsorbate system. The adsorbate properties determining the sorption capacity are following:

- concentration level,
- chemical forms,
- water solubility,
- valency number,
- hydration energy.

The concentration level is important when deciding on the best adsorbent but without knowledge of the form of heavy metals in water, treatment may be ineffective. Heavy metals can be present in drinking water in several different chemical forms and therefore, testing of the water is essential before choosing adsorbent. All heavy metals exist in waters in colloidal, particulate, and dissolved phases, although dissolved concentrations are generally low. The colloidal and particulate metal may be found in: hydroxides, oxides, silicates, or sulfides or adsorbed to clay, silica, or organic matter. The soluble forms are generally ions or unionized organometallic chelates or complexes. Organic compounds form strong complexes with most metals in aquatic systems.

The solubility of a compound in water tends to be inversely proportional to the amount of sorption that the contaminant can undergo. Adsorption of elements is also dependent on their valency and hydration energy.

The sorption of heavy metals in water is also controlled by reaction environment characteristics, such as: pH, redox potential, ionic strength and competing ions:

(1) The pH can affect sorption considerably because it can affect the solubility/precipitation of a compound. Certain compounds dissolve better under certain pH's. A lower pH increases the competition between metal and hydrogen ions for binding sites. A decrease in pH may also dissolve metal-carbonate complexes, releasing free metal ions into the water column.
(2) A decreased redox potential, as is often seen under oxygen deficient conditions, will change the composition of metal complexes and release the metal ions into the overlying water.
(3) Ionic strength can influence on the adsorption processes. Salinity increase competition between cations and metals for binding sites.
(4) The presence of some competing ions could inhibit the sorption of contaminants such as heavy metals and as a result reduce the efficiency of metal removal.
(5) The length of contact time between the water and the adsorbent material, governed by the rate of water flow and the amount/volume of adsorbent, has a significant effect on adsorption of contaminants. More contact time results in greater adsorption until the adsorption reaction reach an equilibrium.

8.3 ADSORPTION TECHNOLOGY

The efficiency of adsorption is mainly determined by the type of adsorbate but also the exposure/adsorption time, a dose (in case of powdery adsorbents), and also the efficiency/result of the preliminary water treatment. The time of adsorbate exposure to the activated bed depends on the type of a bed and temperature of water (at lower temperature the exposure time should be longer/extended) (Nawrocki & Biłzor, 2000).

The application of adsorbates and the selection of technical solutions allowing for the contact of an adsorbate with treated water are affected by the form of adsorbate. Powdery adsorbents can be added to raw water during coagulation (which reduces the doses of adsorbents) or before filtration. They can also be added to the high rotation stirred chambers or before settling tanks with a suspended bed. In the first case, the addition of an adsorbate later in the process than the coagulant is justified due to the fact that during simultaneous dosing the adsorbent particles are built in the structure of forming floccules. As a result, the adsorption capacity of the adsorbent can be reduced. The addition of an adsorbent improves the sedimentation properties of agglomerates formed during the process of flocculation.

Grainy or granular adsorbents can be applied in flow-through systems as a layer in multilayer filtration beds or sorption beds. From the economical point of view, the addition of an adsorbent to raw water is rationalized only when adsorbent is added periodically. During the continuous application the required doses of an adsorbent are higher than in other technological solutions. In order to reduce the dose of an adsorbent and to extend the exposure time/contact time of water and an adsorbent, it is recommended to identify the best dosing point for each syststem. The process can be conducted in a contact chamber (this increases the number of equipment needed for the process) or the adsorbent can be dosed directly to a pipeline system to the filters. In this solution the adsorbent remains in the filtration bed during the entire filtration cycle (Kowal & Świderska-Bróż, 2000).

8.4 APPLICATIONS OF ADSORBENT MATERIALS FOR METALS REMOVAL FROM WATER

The choice of an adsorbent for heavy metals removal from drinking water must be guided by some criteria: metal adsorption must be thermodynamically possible and the process must be fast.

8.4.1 Zeolites

Zeolites are naturally occurring structured minerals with high cation exchange and ion adsorption capacity. The zeolite crystal pore network is assumed to consist of transport pores, which offer the sorbate access to the crystal interior, and short pores intersecting the transport pores, which act as a capacity sink, with the sorbate in each of these pore types being regarded as distinct sorbed species or phases. They have a solid microporous structure with approximately 50% of intrinsic void space with the surface up to 1500 m^2/g. The size of cavity dimension (i.e. canals and pores) in the zeolite structure is so significant that cations and water molecules can migrate due to the effects of physical and chemical factors. The contaminants can be adsorbed in the pores of the aluminosilicate skeleton. The size and shape of intrinsic canals determines which cations and molecules can enter or remain outside the crystal structure. Water is returned and again absorbed by the zeolite (Bień & Zabochnicka-Świątek, 2009).

Aluminosilicates such as clinoptilolite could be used for metals (As, Cu, Pb, Ni) removal from water. The most important properties of zeolites are as follows (Zabochnicka-Świątek, 2007):

- adsorption ability of different capacity for a number of compounds (including metal-ions, vapours and gases),
- high chemical-, temperature- and radiation stability,
- low density and large void volume of dehydrated samples,
- reversible water absorption.

Adsorption properties of zeolites towards metal ions are related to the presence of micropores which have a densely arranged structure. As adsorbents zeolites show adsorption selectivity, molecular-sieves properties and high thermal stability (Zabochnicka-Świątek *et al.* 2008).

Sorption capacity measurement provides most direct ways of characterizing a zeolite sample. However, the information derived from capacity measurement generally provides only an estimate of sample purity and/or evidence of consistency with a known structure, rather than a means of differentiation between different structures.

Sorption of cations by clinoptilolite from contaminated environment is complex. The size and shape of pores in zeolite, ions charge, solution strength, pH and temperature influence on the sorption and ion exchange selectivity. Adsorption of elements in the zeolite pores and skeletal canals is strongly dependent on their valency and hydration energy. The lower valency number and hydration energy the process of sorption is more effective (Zabochnicka-Świątek & Stępniak).

Many investigators showed that various types of zeolites can be produced from solid by-product. Zeolites synthesized from fly ash have many potential applications in environmental protection. The type of fly ash used in the synthesis, kind of the method applied, temperature and solution/fly ash ratio determine the type of zeolite of different structure and efficiency. Modifications of the conventional hydrothermal technique (mixing of fly ash with NaOH solution) for synthesize of zeolites optimise the selectivity and sorption efficiency. Zeolites could be synthesized from other products for example, solid by-product of oil shale (Bień & Zabochnicka-Świątek, 2009). Synthetic zeolites are useful in heavy metal removal because of their controlled and known physico-chemical properties relative to that for natural zeolites (Shevade, R. G. Ford, 2004).

The effectiveness of metals removal by zeolites depends on the concentration of ions in solution and type of the metal ions to be removed. The zeolitic material must be added in sufficient amount to adsorb the all contents of heavy metal and in particular stage in water treatment.

Heavy metals can be removed by application of biosorbents produced on the basis of the zeolite and algae thus providing an economically feasible technology for water treatment. Microbial biomass can passively bind large amounts of metal – biosorption. Biosorption is possible by both living and nonliving biomass. The living biomass is also able to sequester metal intracellularly by an active process – bioaccumulation. The process of biosorption has many attractive features including the selective removal of metal(s) over a broad range of pH and temperature, its rapid kinetics of adsorption and desorption (Vijayaraghavan *et al.* 2004).

8.4.2 Activated carbon

Activated carbon (AC) is a natural material derived from bituminous coal, lignite, wood, coconut shell and so on, activated by steam and other means, and each one has different adsorption properties. Most popular forms of activated carbon used in the treatment of drinking water are granular activated carbon (GAC), extruded solid carbon block (CB) and powdered activated carbon (PAC). Its specific physicochemical properties allow it to function not only as an adsorbent, but also as a catalyst and catalyst carrier (Okoniewska & Zabochnicka-Świątek, 2008). Activated carbon has a surface area approximately 1000–1500 m^2/g.

Activated carbon has a great potential for sorption because of its well-developed surface area. The removal efficiency is influenced by many factors:

- particle size,
- chemical nature of AC,
- adsorbent modification procedure,
- type and concentration of the metal ions,
- the temperature and pH of the water,
- the flow rate or time exposure of water to AC.

Methyl mercury/methylated mercury can be the most efficiently removed by AC among heavy metals. Activated carbon can also change the oxidation state of metal ions resulting in their precipitation. It is confirmed that the activated carbon catalyzed the oxidation of Fe(II)–Fe(III) and also caused the reduction of Cr(VI)–Cr(III). The oxides of these metals are retained by the filtration.

The most common/typical method for adsorption on activated carbon is combining the adsorption with coagulation. In case of powdery AC, the increase in a dose of powdery adsorbant allows for reducing the time of exposure to water. The adsorption on grainy or granular AC can be combined with the process of filtration by using carbon as a layer in a multilayer filtration bed or separately in the sorption columns with a fixed or suspended bed. All suspended and colloidal contaminants should be removed before water is exposed to the adsorbent (Kowal & Świderska-Bróż, 2000).

Activated carbon constitutes a good medium for the growth of microorganisms and at the same time is efficient in removal of chlorine. Consequently, water subjected to sorption on activated carbon has to be disinfected.

8.4.3 Biosorbents

In recent years, many low-cost sorbents have been investigated, but the biosorbents have since proven to be the most effective and promising substrates. Biosorption is the removal of heavy metals and other hazardous substances by the passive binding to non-living micro-organisms (algae, fungi and bacteria) and other biomass (peat, rice hull, wheat shell, fruit peel, leaves, saw dust, bark of trees, macrofungus etc.) from an aqueous solutions. Contrary to synthetic sorbents the biosorbents are bio-renewable, biodegradable and cost-effective. As a rule, biosorbents are distinguished for high sorption selectivity for heavy metals. The alkaline and alkaline earth metals are not sorbed. Therefore, the biosorbents are most suitable for treatment of drinking water, as they remove the hazardous substances only; meanwhile the essential minerals (calcium, magnesium) remain in water (Davis *et al.* 2003; Davis *et al.* 2000; Sari & Tuzen, 2009; Pehlivan *et al.* 2009; Rao & Khan, 2009; Atkinson *et al.* 1998).

Among the number of biosorbents investigated, algae and chitin containing substances have been proven to be the most effective and promising substrates for heavy metal removal. These biosorbents contain in chain the functional groups such as amine, amide, carboxylic, hydroxyl, which are responsible for heavy metal chelation. The complex formation occurs through the surface groups functioning as ligands. The lone pair of electrons present in the nitrogen or oxygen of functional groups can establish dative bonds forming surface complexes with transition metal ions. These metals, as a rule, act as hazardous substances for living organisms.

Chitin and its deacetylated derivative chitosan are produced from seawater and freshwater crustaceans (crab, krill, shrimp etc). Chitosan is able to sorb both heavy metals and organic compounds. The sorption ability of chitosan in many cases exceeds that of synthetic sorbents. The regularities of heavy metal sorption by chitosan are widely studied. The chitosan sorption ability depends on its physical and chemical properties. The deacetylation degree (the number of amine groups in polymer molecule) has the decisive influence.

In aqueous solutions chitin is protonated by a hydrogen ion. The protonated and unprotonated amine of chitosan interacts with the metal ion giving a complex compound.

$$\text{Chit-NH}_2 + \text{H}_3\text{O}^+ \longleftrightarrow \text{Chit-NH}_3^+ + \text{H}_2\text{O};$$

$$\text{Chit-NH}_2 + \text{M}^{n+} \longrightarrow (\text{Complexes})^{n+}$$

$$\text{Chit-NH}_3^+ + \text{M}^{n+} + \text{H}_2\text{O} \longrightarrow (\text{Complex})^{n+} + \text{H}_3\text{O}^+$$

The protonated chitosan can also electrostatically interact with negatively charged material, thus removing dangerous anions from water. It has been demonstrated, that the adsorption of nitrate by chitosan hydrobeads proceeds in the pH range close to neutral. The adsorption process was found to be temperature dependant with an optimum activity at 30°C. The protonated amine groups interact also with arsenite, arsenate, fluoride anions and so on (Chatterjee & Woo, 2009; Sundaram *et al.* 2009).

There exists the number of other biosorbents, which are capable to remove also other hazardous substances from water. Based upon the biochemical mechanisms that explain arsenic toxicity, Teixeira and Ciminelli (2005) proposed a waste biomass with a high fibrous protein content obtained from chicken feathers, which can be used for selective As(III) adsorption. Prior to adsorption, the disulfide bridges present in the biomass are reduced by thioglycolate.

However, the sorption ability of biosorbents usually is low; their regeneration in many cases still remains to be solved. In order to enhance the sorption rate and pollution removal efficiency the different composite materials with biosorbents are exploited. The U.S. Bureau of Mines has developed porous beads containing immobilized biological materials for removing metal contaminants from waste waters. The beads, designated as BIO-FIX beads, are prepared by blending biomass, such as sphagnum peat moss or algae, into a polymer solution and spraying the mixture into water.

The ability of biosorbents to remove the hazardous substances only from water makes the biosorption an attractive technology for the preparing of drinking water. For the most part biosorption is used in rural and remote areas as separate surface or groundwater treatment system in order to prepare the water suitable for food. The filters containing biosorbents such as ARGO (Russia) are used for preparing of drinking water from polluted with heavy metals surface waters. These filters are also used for treatment of water from wells. ARGO filters are capable during the year to prepare ~7 m^3 of drinking water. The biosorbents containing filters ECOLAN (Ukraine) are used for treatment and protection of surface aquifers used as a source for drinking water in large scale.

The biosorbents such as a peat, sawdust, straw and other agriculture wastes are used for cleaning of drinking water sources in emergency cases, when these sources are polluted with untreated industrial or agricultural wastewaters.

8.4.4 Iron oxides

Iron oxides are naturally occurring secondary minerals that consist of Fe(III) (sometimes Fe(II)) coordinated octahedrally to oxygen and/or hydroxyl-groups. Various types of iron oxides are found. They differ in the way the octahedral is arranged in space and thus in crystallinity. Iron oxides has a great potential for sorption because of their large surface area and the presence reactive OH–surface groups. The more amorphous the iron oxide is, the larger surface area and the better sorption capacity. The surface area for iron oxides are in the range of 8–350 m^2/g, with ferrihydrite being the most amorphous (Jambor & Dutrizac, 1998; Cornell *et al.* 2003). In aqueous environments iron oxides have pH-dependent charges, because of the protonation and de-protonation of the functional groups on the surface. Generally they are negatively charged under alkaline conditions and positively charged in acidic conditions (Appelo & Postma, 2005). Therefore capable of adsorbing both cations (e.g. Zn^{2+}, Cu^{2+} and Pb^{2+}) and anions (e.g. $CrO_4{}^{2-}$, $AsO_4{}^{3-}$) depending on pH.

$$\equiv FeOH + H^+ \longleftrightarrow \equiv FeOH_2{}^+$$
$$\equiv FeOH + OH^- \longleftrightarrow \equiv FeO^- + H_2O$$

Traditional treatment of water with iron oxides has relied on the addition of an Fe(II) salt to the water – followed by oxidation in neutral to slightly alkaline solutions, and thereby a precipitation of Fe(III) as ferrihydrite (Cornell and Schwertmann, 2003). During precipitation metals are built in to the iron oxide structure and after the precipitated ferrihydrite can function as an adsorbent. However this process generates large amounts of hydrous ferrihydrite sludge and is expensive therefore the use of other types of iron oxides as sorbents has recently become a focus area.

Granular ferric hydroxide (Asgari *et al.* 2008), zero-valent iron, goethite (Mohan & Pitmann, 2006) and different types of sorbent materials such as sand, cement and polymeric materials coated with iron oxides (Benjamin *et al.* 1996; Genc-Fuhrman *et al.* 2007; Katsoyiannis & Zoubolis, 2002; Khaodhiar *et al.* 1998; Kundu & Gupta, 2005) are among the iron based sorbents that either used or being examined for future use. An important issue with these iron oxides is to maintain good hydraulic conditions.

8.5 ADVANTAGES AND DISADVANTAGES OF ADSORTPION

	Adsorption
Advantages	• utilized adsorbent is removed from the water treatment system during rinsing of the filtration beds, this reduces the risk of desorption of the previously sorbed contaminants, • usage of powdery adsorbents enables to correct the dose of an adsorbent depending on the concentration of adsorbates in treated water which allows for periodical application of the adsorption process.
Disadvantages	• adsorbent particles kept in the filtration bed increase the pressure drop across the bed, • there are no practical options for the recovery of used powdery adsorbents, • there is a potential risk of transferring small adsorbent–adsorbate agglomerates into the treated water, • biosorption due to biodegradability of sorbents used is more complicated water treatment technology when sorption onto mineral or synthetic biosorbents, • the sorption ability of biosorbents is low.

The empty bed contact time is the definition of the time the water that needs treatment is in contact with the adsorbent, and is therefore the main design parameter. The high efficiency of adsorption is achieved in the sorption column with beds of increased heights (even up to 3.0 m). At optimal filtration velocity, this allows for longer exposure time and better exploitation of adsorption capacity. This exposure time, also termed the empty bed contact time (EBCT) is the main design parameter. The velocity of water flow through a filtration and adsorption bed is limited by the filtration velocity and usually amounts to 10 m/h. For the sorption columns the velocity of water flow ranges from 5 to 30 m/h.

	Sorption columns
Advantages	• a new adsorbent can be added in the upper part of the column and the used adsorbent is removed without the necessity of shutting down the column • the application of sorption columns in a multicolumn serial system allows for adding and regenerating the columns, • the serial column systems use less adsorbent than the single-stage column systems.
Disadvantages	• the efficiency of adsorption strongly depends on the competitiveness of contaminants, • the occurrence of competing ions in water can result in desorption of previously adsorbed substances.

KEY POINTS

Adsorbents could be used after flocculation and sedimentation, during the filtration process (granular) or during the coagulation process (powdered). In order to increase contact time, adsorbents should be added before filtration – but it needs additional installation. The use of adsorption column is the most efficient way.

Zeolites are the best natural filter medium available for treatment of water for metals removal. It offers superior performance to sand and carbon filters, giving purer water and higher throughput rates with less maintenance required. The application of zeolites provide an economical means of removing mixed heavy metals from water. Physical and/or chemical regeneration of used zeolitic adsorbents facilitate the protection of natural environment, provided that the used regenerant can be disposed of safely.

Activated carbons are non-selective sorbents and presence of the competing ions could reduce the efficiency of metal removal. All activated carbon filters, do not naturally reduce the levels of soluble salts (including nitrates), fluoride, and some other potentially harmful minerals like arsenic (unless specially designed) and cadmium.

REFERENCES

Appelo C. A. J. and Postma D. (2005). Geochemistry, Groundwater and Pollution. A.A. Balkema Publishers, Amsterdam.

Asgari A. R., Vaezi F., Nasseri S., Dördelmann O., Mahvi A. H. and Fard E. H. (2008). Removal of hexavalent chromium from drinking water by granular ferric hydroxide. *Iran. J. Environ. Health Sci. Eng.*, **5**, 277–282.

Atkinson B. W., Bux F. and Kasan H. C. (1998). Considerations for application of biosorption technology to remediate metal-contaminated industrial effluents. *Water SA.*, **24**(2), 129–136.

Bień J. B. and Zabochnicka-Świątek M. (2007). Ion exchange selectivity and adsorption capacity of clinoptilolite. In: Environmental Protection into the Future, W. Nowak and J. B. Bień (eds), Wydawnictwo Politechniki Częstochowskiej, Częstochowa, Poland, pp. 383–393.

Benjamin M. M., Sletten R. S., Bailey R. P. and Bennett T. (1996). Sorption and filtration of metals using iron-oxide-coated sand. *Water Res.*, **30**, 2609–2620.

Chatterjee S. and Woo S. H. (2009). The removal of nitrate from aqueous solutions by chitosan hydrogel beads. *J. Hazard. Mater.*, **164**, 1012–1018.

Cornell R. M. and Schwertmann U. (2003). The Iron Oxides: Structure, Properties, Reactions, Occurences and Uses. Wiley-VCH, Weinheim, Germany.

Davis T. A., Volesky B. and Mucci A. (2003). A review of the biochemistry of heavy metal biosorption by brown algae. Review. *Water Res.*, **37**, 4311–4330.

Davis T. A., Volesky B. and Vieira R. H. S. F. (2000). Sargassum seaweed as biosorbent for heavy metals. *Water Res.*, **34**(17), 4270–4278.

Genc-Furhman H., Mikkelsen P. S. and Ledin A. (2007). Simultaneous removal of As, Cd, Cr, Cu, Ni and Zn from stormwater: experimental comparison of 11 different sorbents. *Water Res.*, **41**, 591–602.

Jambor J. L. and Dutrizac J. E. (1998). Occurrence and constitution of natural and synthetic ferrihydrite, a widespread iron oxyhydroxide. *Chem. Rev.*, **7**, 2549–2585.

Katsoyiannis I. A. and Zouboulis A. I. (2002). Removal of arsenic from contaminated water sources by sorption onto iron-oxide-coated polymeric materials. *Water Res.*, **36**, 5141–5155.

Khaodhiar S., Azizian M. F. and Nelson P. O. (1998). Equilibrium modeling of arsenic, chromium, and copper adsorption on an iron-oxide-coated sand, *Abstr. Papers Am. Chem. Soc.* **215**, U581.

Kowal A. L. and Świderska-Bróż M. (2000). Oczyszczanie Wody (Water purification). Wydawnictwo naukowe PWN, Warszawa-Wrocław.

Kundu S. and Gupta A. K. (2005). Sorption kinetics of As(V) with iron-oxide-coated cement – a new adsorbent and its application in the removal of arsenic from real-life groundwater samples. *J. Environ. Sci. Health Part A: Toxic/Hazard. Subs. Environ. Eng.*, **40**, 2227–2246.

Mohan D. and Pitmann C. U. (2006). Activated carbons and low cost adsorbents for remediation of tri- and hexavelent chromium from water. *J. Hazard. Mater.*, **137**, 762–811.

Nawrocki J. and Biłozor S. (2000). Uzdatniaie Wody. Procesy Chemiczne i Biologiczne (Water treatment. Chemical and biological processes). Wydawnictwo naukowe PWN, Warszawa Poznań.

Okoniewska E. and Zabochnicka-Świątek M. (2008). The influence of modification of activated carbon on organic manganese removal. 2nd International Conference Metals and related substances in drinking water, Cost action 637, Lisbon, Portugal, 29–31 October 2008.

Pehlivan E., Altun T., Cetin S. and Bhanger M. I. (2009). Lead sorption by waste biomass of hazelnut and almond shell. *J. Hazard. Mater.*, **167**, 1203–1208.

Rao R. A. K. and Khan M. A. (2009). Biosorption of bivalent metal ions from aqueous solution by an agricultural waste: Kinetics, thermodynamics and environmental effects. *Colloids and Surfaces A: Physicochem. Eng. Aspects*, **332**, 121–128.

Sarı A. and Tuzen M. (2009). Kinetic and equilibrium studies of biosorption of Pb(II) and Cd(II) from aqueous solution by macrofungus (Amanita rubescens) biomass. *J. Hazard. Mater.*, **164**, 1004–1011.

Shevade S. and Ford R. G. (2004). Use of synthetic zeolites for arsenate removal from pollutant water. *Water Res.*, **38**, 3197–3204.

Sundaram C. S., Viswanathan N. and Meenakshi S. (2009). Defluoridation of water using magnesia/chitosan composite. *J. Hazard. Mater.*, **163**, 618–624.

Teixeira M. C. and Ciminelli V. T. (2005). Development of a biosorbent for arsenite: structural modeling based on x-ray spectroscopy. *Environ. Sci. Technol.*, **39**, 895–900.

Vijayaraghavan K., Jegan J. R., Palanivelu K. and Elan M. (2004). Copper removal from aqueous solution by marine green alga Ulva reticulate. *Electron. J. Biotechnol.*, **7**(1), 47–54. ISSN: 0717-3458.

Zabochnicka-Świątek M. (2007). Czynniki wpływające na pojemność adsorpcyjną i selektywność jonowymienną klinoptylolitu wobec kationów metali ciężkich, Inżynieria i Ochrona Środowiska (Factors affecting the adsorption capacity and ion exchange selectivity of clinoptilolite towards heavy metal cations). *Częstochowa*, **10**(1), 27–43.

Zabochnicka-Świątek M., Stańczyk-Mazanek E. and Bień J. B. (2008). Immobilization of Cu^{2+} and Zn^{2+} in the presence of Ba^{2+} and Sr^{2+} by clinoptilolite. *Polish J. Environ. Stud.*, **17**(3A), 605–610.

Zabochnicka-Świątek M. and Stępniak L. (2008). The potential applications of aluminosilicates for metals removal from water. 2nd International Conference Metals and related substances in drinking water, Cost action 637, Lisbon, Portugal, 29–31 October 2008.

Chapter 9

Ion exchange processes

Magdalena Zabochnicka-Swiatek

9.1 INTRODUCTION

Ion exchange is a reversible chemical reaction. Ion (an atom or molecule that has lost or gained an electron and thus acquired an electrical charge) from solution is exchanged for a similarly charged ion attached to an immobile solid particle. These solid ion exchange particles are either naturally occurring inorganic adsorbents or synthetically produced organic resins. The synthetic organic resins are the predominant type used today because their characteristics can be tailored to specific applications.

With reference to the type of ions subjected to the exchange, the ionites are divided into:

- Cation exchangers (CE) – they exchange cations,
- Anion exchangers (AE) – they exchange anions.

Cation exchangers show the character of acids or their salts and anion exchangers show the character of bases or their salts. Bipolar ionites comprise both acid and base groups, and also ampholytes which have these groups as well. Depending on the pH, the ampholytes can function as acid or base groups.

9.2 FACTORS INFLUENCING ION EXCHANGE SELECTIVITY

Ion exchange enable removal of all ions from the solution or only selected ions can be removed. Selectivity is defined as a selective exchange of a given counter-ion in the presence of other counter-ions. As a result, the quantitative ratio of these two types of ions subjected to the exchange after reaching the equilibrium is different in the exchanger than in the solution.

The main parameters of the ion exchange process are: the physical and chemical composition of solutions subjected to treatment, the concentration of electrolytes, the type of exchanged ions, the quantity of electric charge and the way of running the process (Zhao *et al.* 2002).

The composition of ionites – ionites should have the adequate porosity and a significant number of channels allowing for the transport of ions. The three-dimensional, the size and number of channels determine the mechanism of ion exchange which can occur inside or on the surface of ionite grains, or can be mixed.

The type of functional groups determines the character of exchange reaction, and the number of groups affects the ionite exchange capacity and the expanding degree. In practice, the excessive increase in the ionite exchange capacity is not feasible as with the increase in the capacity the expanding degree also increases, which in consequence reduces the durability of ionite.

The type of ions subjected to exchange – the hydrodynamic radius and electric charge of ions influence their exchangeable energy which in consequence affects the mechanism of ion exchange. The higher the energy, the higher the electric charge and the ion atomic mass and the smaller its hydrodynamic radius are.

The energy required for ions to enter the ionite depends on the concentration and dissociation degree of electrolyte, pH, temperature and the type of ionite. This means that the values of exchangeable energy of the same ions can be different for various ionites.

The pH of the treated solutions determines the mechanisms of dissociation of functional groups which directly influences the dynamics and efficiency of the process.

The concentration of the electrolyte subjected to the treatment influences the exchange of ions of different valences, and with diluting the solution the exchange of ions with higher electric charge increases.

The efficiency of ion exchange increases with the increase in the temperature of the solutions subjected to treatment/treated solutions.

The complete operation cycle for ion exchangers consists of:

- The ionite operation time required to reach the breakthrough point,
- The regeneration of a resin bed which requires subsequent rinsing.

Apart from the direct application of ionites accordingly to the form of metals present in water, the modification of these metal forms to for example anions, and their removal during deanionation, is also possible at the earlier stages of the process (before the ion exchange) (Zhao *et al.* 2002).

The two-ionite beds (i.e. mixed beds) are recommended for treatment of water with cation and anion forms of contaminants as they assure simultaneous decationation and deanionation.

Anionites are also very efficient in the exchange of uranium anion forms [$UO_2(CO_3)_{2-}$ and $UO_2(CO_3)_{3-}$] which are present in natural water. In this case, the exchange on highly alkaline anionites operating in a chloride cycle is the most efficient solution (99–100%).

9.3 APPLICATIONS OF ION EXCHANGE MATERIALS FOR METALS REMOVAL FROM WATER

9.3.1 Zeolites

The ion exchange selectivity of natural clinoptilolite for inorganic cations has been investigated by many authors. Selective cation exchange may provide an economical means of removing mixed heavy metals from effluents.

Cations present in the canals in the zeolite structure readily undergo the process of exchange (exchangeable cations). Ions of heavy metals can be replaced with Na, Ca, Mg and K. Si and Al ions in normal conditions do not undergo any exchange processes (tetrahedronic or skeletal atoms) (Bień & Zabochnicka-Świątek, 2007). The elements can undergo ion exchange processes on zeolite according to the following reaction:

$$Me^{+} + Na[zeolite] \longleftrightarrow Me[zeolite] + Na^{+}$$

Argun (2008) stated that the use of zeolites as ion-exchanger for environmental protection and other applications has been stimulated by good results obtained in testing and by the non-toxic nature of these materials. According to Chughtai *et al.* [3] the process of ion exchange refers to the replacement of toxic heavy metals ions in solution by the more benign counter-ions that balance the surface charge of the solid exchanger.

Clinoptilolite tends to elevate the solution acidity. Acidity of solutions is expected to change during ion exchange, as H^+ ions are exchanged with the cations initially present in zeolite structure. Therefore, during the ion-exchange process, acidity is changing due to H^+ uptake by clinoptilolite, but also due to the uptake of metal cations, which have a tendency to give acidic solutions. As a result, the competitive uptake of H^+ cations is considered to be the reason of lower uptake of metals in more acidic environment (Inglezakis *et al.* 2007).

According to the literature, modifications of zeolites enhance their ion exchange ability by optimising the selectivity and sorption efficiency. There are a number of different ways that zeolites can be modified, for example by physical and chemical pretreatment. The important method to improve their ion-exchange capacity could be pretreatment them by acids, bases and surfactants, and so on. Practically, any pretreatment operation increases the content in a single cation, what is called homoionic form (Gunay *et al.* 2007).

Chughtai & Keane (1998) examined heavy-metal ion exchange from nickel/copper mixtures from aqueous solution and regeneration of the used zeolite by back exchange. He found that copper removal was much greater than that of nickel for all the zeolite exchangers under identical experimental conditions.

Various methods of lead removal from water have been applied including chemical precipitation, membrane processes, ion-exchange and adsorption.

The most efficient media able to remove lead from water are zeolites, natural or synthetic ones. The choice of an adsorbent for lead removal from drinking water must be guided by at least three criteria: lead adsorption must be thermodynamically possible, kinetic limitations must not prevent it even with a very low contact time and the adsorbent must be authorized for water contact and not lead to bacteria growth (Inglezakis, 2007).

Gunay *et al.* (2007) also investigated sorption of lead. The investigators observed that clinoptilolite is an effective adsorbent for the removal of Pb(II) from aqueous solution. The batch studies indicated that the lead adsorption on clinoptilolite increased with increase in initial Pb(II) concentration. The experimental maximum adsorption capacities of Pb(II) onto clinoptilolite were 80.933 and 122.400 mg/g for raw and pretreated clinoptilolite, respectively, for the initial concentration of 400 mg/L. Adsorption of Pb(II) appears to take place by an ion-exchange mechanism.

Inglezakis *et al.* (2007) studied removal of Pb(II) from aqueous solutions by using clinoptilolite and bentonite as adsorbents. The researchers stated that clinoptilolite dust is found to be more efficient than granular clinoptilolite. Agitation and temperature affected the uptake of Pb(II), especially in the case of granular clinoptilolite (2.5–5.0 mm). Clinoptilolite dust seems to be more effective than clinoptilolite 2.5–5.0 mm at 45°C, but at 60°C it seems to reach the same level of removal (50%). The highest removal level reached by clinoptilolite was 55%. Finally, they concluded that acidity of the aqueous solution influences the removal of lead by the minerale. The adsorption of lead increases with an increase in pH of the solution from 1 to 4.

Heavy metals pollution is usually found as mixed metals. In such cases competition processes influence the sorption efficiency.

Mondale *et al.* (1995) determined the selectivity series of clinoptilolite for removal of mixed heavy metals from water.

$$Pb^{2+} > Cd^{2+} > Zn^{2+} \geq Cu^{2+} > Ni^{2+} > Hg^{2+}$$

The authors show that the sorption of Pb^{2+} by the clinoptilolite is the most effective.

Inglezakis *et al.* (2003) investigated the ion exchange of Pb^{2+}, Cu^{2+}, Fe^{3+}, and Cr^{3+} on natural clinoptilolite. The researchers found that the selectivity of clinoptilolite for metal cations is dependent on the initial acidity of the metal solutions. In two-component and four-component solutions, selectivity is following the same order:

$$Pb^{2+} > Fe^{3+} > Cr^{3+} \geq Cu^{2+}$$

as measured for single metal solutions at the same acidity pH $= 2$, except for the couple Cu^{2+}– Cr^{3+}. This order is established from the first days of equilibration, but the selectivity ratio of metals is changing as ion exchange proceeds to reach equilibrium. In multiple-cation solution these changes during exchange could be explained by the fact that ion exchange is a dynamic process where the binding forces between cations and the zeolite matrix are relatively weak. They found that selectivity in single metal solutions where acidity is not adjusted is following the order:

$$Pb^{2+} > Cr^{3+} > Fe^{3+} \approx Cu^{2+}$$

The researcher concluded that selectivity ratio is not constant during ion exchange and is concentration-dependent (Inglezakis, 2007).

Sprynsky *et al.* (2006) studied selection mechanism of heavy metal (Pb^{2+}, Cu^{2+}, Ni^{2+} and Cd^{2+}) adsorption on clinoptilolite and also found the the adsorption decrease in the more acidic medium. The investigators reported that adsorption of lead, copper, cadmium, and nickel onto clinoptilolite has an ion – exchange nature.

Nearly 40% of copper and cadmium and nearly 90% of nickel are sorbed during the first stage on the microcrystal's surface. Efficiency of metals removal from solutions by the clinoptilolite is inversely proportional to the metal concentration and the metals form the following order for adsorption efficiency:

$$Pb^{2+} > Cu^{2+} > Cd^{2+} > Ni^{2+}$$

The authors stated that the slight difference between adsorption capacity of the clinoptilolite toward lead, copper, and cadmium from single and multicomponent solutions may tesify to individual sorption centers of the zeolite for each of these metals. Decrease of nickel adsorption from multicomponent solutions is probably caused by competition processes (Sprynsky *et al.* 2006).

Chughtai *et al.* (1998) investigated the removal of lead from aqueous solution by ion exchange with synthetic zeolite. The researchers concluded that synthetic aluminosilicate zeolites act as efficient porous solid exchange media:

$$Pb^{2+} > Cd^{2+} > Cu^{2+} > Ni^{2+}$$

Lead removal was much greater than that of cadmium under identical experimental conditions.

9.3.2 Organic and inorganic ion exchangers

The ion exchangers can be applied for a selective elimination of heavy metals from drinking water. Ion exchangers with iminodiacetic acid groups are applied in industrial water treatment. The selectivity effect is caused by the presence of nitrogen atom in the functional groups. For drinking water treatment, there is a need to pre-load them with calcium ions to avoid the elimination of calcium ions (Zhao *et al.* 2002).

Weakly basic exchangers are allowed for food and drinking water treatment. The main advantages is that they cannot adsorb alkaline and alkaline earth ions.

A sodium titanate ion exchangers is an inorganic ion exchanger that is efficient in removing of transition metals. An organic ion exchangers is represented by aminophosphonate. Both of them are several times more efficient than conventional resins in removing toxic and harmful transition metals, such as: Cu, Zn, Mn and Ni (Vaaramaa & Lehto, 2003).

Polystyrene ion exchangers is effective for removal of heavy metal ions, such as Cd and Pb in the presence of the complexing agent such as EDTA. During the process the influence of ionic strength, pH and solution interface should be monitored (Kołodyńska *et al.* 2009).

For removal of Pb was applied: cation exchangers with functional phosphonic or aminophosphonic groups, ion exchangers obtained from modification of gel type copolymer consisting of vinylbenzene chloride, styrene, and divinylbenzene, and also organic ligands: aminopolycarboxylic acids (e.g. EDTA) (Dąbrowski *et al.* 2004).

Polystyrenesulphonic cation exchanger Dowex 50W-X4, chelating ion exchangers containing phosphonic and sulphonic groups, polystyrenedivinylbenzene, chelating resin prepared on the basis of copolymer of acrylnitrile/ethyl acrylate/divinylbenzene can be applied for removal of Cd.

Polystyrene-sulphone, carboxylic, iminodiacetate Chelex-100 ion exchangers can be used for removal of Ni.

For removal of Cu from less acidic solutions can be used: chelating ion exchanger Dowex XFS-4196 which can be regenerated by means of sulphuric acid. Duolite ES-346 with polystyrenedivinylbenzene skeleton, the chelating phenol formaldehyde resin, amphoteric ion exchange fibres are also widely applied.

9.4 ION EXCHANGE TECHNOLOGY

A pressure tank is the primary system used for the ion exchange process. Pressure columns are used for the process at the working pressure of 0.6 MPa. The draining system is located at the bottom of a column whereas at the top of the column there is the separating and storing system allowing for uniform separation and reception of water and the water regeneration agent used for aeration and rinsing the ionite bed. The ionite bed is placed on the supporting gravel layer with the particle size larger than the openings of the draining system. The height of the cylindrical part of the tank has to assure the required height of the supporting layer and ionite during the operation and aeration of the ionite bed. In case of the columns with a mixed bed (i.e. two-layered bed) the height of the cylindrical part should be larger. Moreover, these columns should be equipped with the additional system for the counter-current regeneration.

In order to design the adequate system, the following parameters should be determined

- The chemical composition/content of water (pH influences the process efficiency),
- Temperature,
- The type of ionite,
- The working capacity of ionite and its suitability for water treatment,
- The permissible hydraulic load of an ionite bed and the linear velocity of water flow through the bed,
- The type and quantity of a regeneration agent,
- The frequency and time required for the regeneration,
- The quantity of water for aeration/fluffing and rinsing the ionite.

The type and unitary usage of the regeneration agent depend on the type and quantity of exchanged ions, the type of ionite, and also temperature of the regeneration solution and requirements for water treated with the ion exchange method (Kowal & Świderska-Bróż, 2000).

	Ion exchange
Advantages	• Selective cation exchange may provide an economical means of removing mixed heavy metals from contaminated water.
Disadvantages	• The capacity of ion exchanger could increase or decrease after the regeneration processes.
	• Regeneration procedure is complicated and there is a need of additional energy.

KEY POINTS

Ion exchange is feasible when an exchanger has a high selectivity for the metal to be removed and the concentration of competing ions is low. Ion exchange processes could be used after flocculation, sedimentation and filtration. Modifications of clinoptilolite enhance its ion exchange ability by optimising the selectivity and sorption efficiency. The differences in the ion exchange selectivity are probably caused by competition processes.

After using, ion exchangers can be cleaned by chemical or thermal processes and reused.

REFERENCES

Argun M. E. (2008). Use of clinoptilolite for the removal of nickel ions from water: kinetics and thermodynamics. *J. Hazard. Mater.*, **150**, 587–595.

Bień J. B. and Zabochnicka-Świątek M. (2007). Ion exchange selectivity and adsorption capacity of clinoptilolite. In: Environmental Protection into the Future, W. Nowak and J. B. Bień (eds), Wydawnictwo Politechniki Częstochowskiej, Częstochowa, Poland, pp. 383–393.

Chughtai A. S. and Keane M. A. (1998). The removal of cadmium and lead from aqueous solution by ion exchange with Na-Y zeolite. *Sep. Purif. Technol.*, **13**, 57–64.

Dąbrowski A., Hubicki Z., Podkościelny P. and Rubens E. (2004). Selective removal of the heavy metal ions from waters and industrial wastewaters by ion-exchange method. *Chemosphere*, **56**, 91–106.

Gunay A., Arslankaya E. and Tosun I. (2007). Lead removal from aqueous solution by natural and pretreated clinoptilolite: adsorption equilibrium and kinetics. *J. Hazard. Mater.*, **146**, 362–371.

Inglezakis V. J., Loizidou M. D. and Grigoropoulou H. P. (2003). Ion exchange of Pb^{2+}, Cu^{2+}, Fe^{3+}, and Cr^{3+} on natural clinoptilolite: selectivity determination and influence of acidity on metal uptake. *J. Colloid Interface Sci.*, **261**, 49–54.

Inglezakis V. J., Stylianou M. A., Gkantzou D. and Loizidou M. D. (2007). Removal of Pb(II) from aqueous solutions by using clinoptilolite and bentonite as adsorbents. *Desalination*, **210**, 248–256.

Kołodyńska D., Skwarek E., Hubicki Z. and Janusz W. (2009). Effect of adsorption of Pb(II) and Cd(II) ions In the presence of EDTA on the chracteristics of electrical double layers at the ion Exchange/NaCl electrolyte solution interface. *J. Colloid Interface Sci.*, **333**, 448–456.

Kowal A. L. and Świderska-Bróż M. (2000). Oczyszczanie Wody (Water purification). Wydawnictwo naukowe PWN, Warszawa-Wrocław.

Mondale K. D., Carland R. M. and Aplan F. F. (1995). The comparative ion exchange capacities of natural sedimentary and synthetic zeolites. *Minerals Eng.*, **8**(4/5), 535–548.

Sprynsky M., Buszewski B., Terzyk A. P. and Namieśnik J. (2006). Study selection mechanism of heavy metal (Pb^{2+}, Cu^{2+}, Ni^{2+} and Cd^{2+}) adsorption on clinoptilolite. *J. Colloid Interface Sci.*, **304**, 21–28.

Vaaramaa K. and Lehto J. (2003). Removal of metals and anions from drinking wter by ion exchange. *Desalination*, **155**, 157–170.

Zhao X., Holl W. H. and Yun G. (2002). Elimination of cadmium trace contaminations from drinking water. *Water Res.*, **36**, 851–858.

Chapter 10

Membrane processes

Asher Brenner and Zdravka Lazarova

10.1 INTRODUCTION

Membrane separation processes are considered emerging technologies capable of improving dramatically traditional processes such as granular filtration, precipitation, disinfection, biological treatment, and to assist in complete elimination of emerging pollutants. Separation of suspended or dissolved constituents from a liquid stream is based on physical characteristics such as size differences or on chemical characteristics such as charge and hydrophobicity of the membrane and the substances to be separated. Membranes are built from synthetic materials which are very thin (less than 1 mm) and have selective permeability. Within the broad family of membrane separation processes there are "loose" membrane-based technologies, the separating mechanism of which is mainly size exclusion. These technologies include Microfiltration (MF) and Ultrafiltration (UF), which are operated successfully under relatively low levels of pressure.

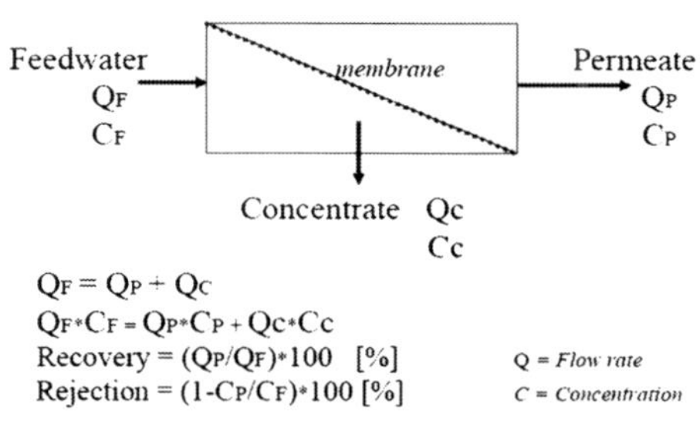

Figure 10.1 Basic definitions of membrane separation processes.

The other type of technologies that can be termed "tight" membrane separation processes require higher levels of pressure to enable free water passage with simultaneous rejection of particulate and even dissolved materials. This group includes Nanofiltration (NF) and Reverse Osmosis (RO). For these two processes, the separation is based not only on physical mechanism related to membrane pore size (relatively small) that serves as a barrier, but also on the chemical structure of membrane material that can dissolve, attract, or reject various substances (Li *et al.* 2008). The later group of processes because of its ability to separate dissolved compounds and ions, can be classified as desalination technology. Basic definitions of flow and mass balance in membrane separation processes are given in Figure 10.1. This can assist process evaluation in regard to suspended solids, salts, or metals removal.

10.2 DESCRIPTION OF TECHNOLOGY

Membrane filtration

MF and UF have gained popularity in recent years in water and wastewater works to substitute the conventional deep-bed filtration for the removal of suspended solids, colloidal particles and microorganisms (especially protozoa and bacteria). In this regard, these processes have been traditionally judged for their efficiency by the turbidity index. However, additional applications include substitution of sedimentation, and of various filtration technologies applied in chemical precipitation processes that result in the formation of solid particles (including metals removal processes). UF which has a smaller pore size than MF can remove partly also organic compounds such as proteins and carbohydrates, and serve as a pretreatment to membrane desalination.

Membrane desalination

Due to the increasing problems of water shortage across the world, there is a growing trend for producing new water sources by desalination of seawater and brackish water. Desalination converts water with high dissolved solids content into water with a very low dissolved solids content, together with the removal of other impurities such as residual organics and microorganisms. Reverse osmosis (RO) is the most common process applied today in desalination of seawater or brackish water for the purpose of drinking water supplies. While osmosis is a natural phenomenon of water diffusion through a semipermeable membrane due to a concentration gradient (the motion is from the low solute concentration to the high solute concentration), reverse osmosis requires input of an external pressure to drive an opposite flow direction.

Nanofiltration (NF) is a more moderate high pressure driven process in which monovalent ions pass freely through the membrane while highly charged multivalent salts are rejected due to the special structure of the NF membrane surface which is negatively charged at neutral and alkaline media. The main advantage of this character is the possibility to achieve high ion rejections at relatively high permeate fluxes, which is very important from practical point of view. Typical NF applications include water softening, desalination of dyestuffs, acid and caustic recovery, and metals removal.

Electrodialysis (ED) belongs also to the desalination processes because of its ability to separate ionic substances (Chapter 7). However, it is motivated by electric voltage and limited to partial rejection of monovalent and divalent salts and of dissociated acids (Van der Bruggen *et al.* 2004).

10.3 IMPLEMENTATION OF TECHNOLOGY FOR THE REMOVAL OF HEAVY METALS AND RELATED SUBSTANCES

Membrane separation processes have been applied widely in water treatment works for a range of applications including the removal of heavy metals and metalloids (Landaburu-Aguirre *et al.* 2006; Mimoune *et al.* 2007; Lee *et al.* 2008; Sang *et al.* 2008). MF and UF technologies are suitable processes for the treatment of suspensions, and can replace the traditional granular filtration. Therefore, they can be applied for separation of particulate forms of various metals following chemical treatment. The tighter membrane separation technologies NF and RO can be used even for the removal of dissolved forms of metals. Of these two processes, NF may be an attractive alternative because it is operated at significantly lower pressures and offers higher fluxes of product water (Košutić *et al.* 2005). In addition to the production of potable water, there are other alternative uses of desalination technologies which include: softening (Nanda *et al.* 2008), natural organic matter (NOM) removal for disinfection by-products (DBP) control (de la Rubia *et al.* 2008), and specific contaminant removal such as heavy metals (Kozlowski & Walkowiak, 2002), radionuclides (Favre-Réguillon *et al.* 2008), or emerging micro-pollutants (Velizarov *et al.* 2008).

10.4 ADVANTAGES AND DISADVANTAGES

Main advantages and disadvantages of membrane separation processes are summarized in Table 10.2. As demonstrated in this table, the main advantage of membrane processes is their high selective separation capabilities that enable selection of a specific high efficient process fitted to the type of contaminants to be removed. On the other hand, the main drawback of this class of technologies is related to membrane material and sensitivity that require tight maintenance, frequent physical and chemical cleaning, and membrane replacement.

Table 10.1 Characteristics of membrane separation processes.

Process	**MF**	**UF**	**NF**	**RO**	**ED**
Driving force	Pressure	Pressure	Pressure	Pressure	Voltage
Operating pressure [bar]	0.5–3	1–5	5–15	10–70	/
Nominal pore size [nm]	10–2000	5–200	1–10	0.1–1	5–200
Substances separated:					
Suspended solids	√	√	√	√	
Colloidal solids	√	√	√	√	
Protozoa	√	√	√	√	
Bacteria	√	√	√	√	
Viruses		√	√	√	
Macromolecules		√	√	√	
Sugars			√	√	
Divalent salts			√	√	√
Monovalent salts				√	√
Dissociated acids			√	√	√
Undissociated acids				√	

1 bar = 10^5 Pascal (Pa) = 10^5 N/m^2 = 14.5 psi [pounds/(square inch)].
1 nm = 10^{-9} m = 10 Angstrom.

Another problem of potential release of metals into water is related to desalinated water, produced for drinking purposes. In this water the levels of alkalinity and essential minerals, such as calcium and magnesium, are very low. Therefore, desalinated water may be associated with inferior taste and corrosion problems that result in the release of metal colloids into water distribution pipes. Tang *et al.* (2006), investigated the impacts of distribution water quality changes caused by blending different source waters on lead release from corrosion in a pilot distribution study. They developed a model which indicated that primary treatment processes such as enhanced coagulation with aluminum sulfate and RO were related to lead release by water quality changes. Therefore, desalinated water is usually stabilized before distribution in order to avoid the problems of corrosion and "red water" incidents in water distribution pipes (Lytle *et al.* 2005). Practices in stabilization typically involve mixing the desalinated water with un-desalinated source water or adding the needed minerals and alkalinity, frequently using limestone dissolution.

Table 10.2 Main advantages and disadvantages of membrane separation processes.

Process	Advantages	Disadvantages
Membrane filtration	• High removal efficiency of suspended & colloidal matter, turbidity, and microorganisms. • Small requirement for process footprint. • Reduced consumption of chemicals.	• Higher energy consumption than the traditional granular filtration. • Relatively high cost of membrane and need to replace it frequently. • Fouling of membranes causing operational problems and cleaning requirements.
Membrane desalination	• A complete barrier to microorganisms. • High removal efficiency of dissolved constituents including metals. • Low footprint.	• High energy consumption. • High cost of membranes. • Requirement for pretreatment to protect membranes. • Fouling of membranes.

10.5 CASE STUDIES

Chilyumova and Thöming (2008), studied filtration of different single model solutions containing mono- and bivalent cations including Ni^{2+} with several NF membranes and determined three model parameters (pore radius, membrane charge and pore dielectric constant) for simulation of a case study of a process integrating nickel recovery. Sang *et al.* (2008) tested a novel nanofiber membrane for the removal of copper, lead and cadmium and found that it can be used for the treatment of groundwater containing these metals with high efficiency. Oehmen *et al.* (2006), studied a novel treatment method for the removal of Arsenic (As) and mercury (Hg^{2+}) by the ion exchange membrane bioreactor (IEMB) process, which incorporates pollutant transport through an ion exchange membrane by Donnan dialysis, with biological removal of the pollutant. As claimed in this study, the IEMB process has a high potential for use in drinking water treatment systems, due to numerous advantages over currently implemented processes, such as minimizing the risk of secondary pollution of the drinking water. Favre-Réguillon *et al.* (2008) showed a high performance of the NF membrane for selective uranium rejection at low pressure (1 bar), illustrating the advantage of NF for the selective removal of uranium from drinking water. Mbarecka *et al.* (2009), showed that the use of ultrafiltration-ion exchange membrane with semi interpenetrating polysulfone and polyacrylic acid network to remove lead, chromium and cadmium from water has shown excellent results mainly at pH superior than 5.7. The high retention was attributed to the complexation between metal ions and carboxylate groups (–COO–) on the inner surface of pores and membrane matrix. Gubbuk *et al.* (2010) showed that transport of the Hg^{2+} through liquid membranes was performed using calix[4]arene nitrile derivatives as a membrane. It was found that calix[4]arene derivatives have the desired ability of selective extraction of Hg^{2+} ions. Thus, this metal ion, which is harmful for nature, can be removed from water even at room temperature and at a low mixing speed (298 K and 100 rpm). The retention of Cu(II) and Ni(II) polyaminocarboxylate complexes in dilute solutions using assisted-UF by polyethylenimine (PEI) and chitosan has been investigated (Zamariotto *et al.* 2010). Metals were very efficiently captured by PEI in the pH-range of 5–8, while high metal retentions were only observed in strong alkaline solutions for chitosan. Polyamines retained metals through acid–base and electrostatic interactions and the better efficiency observed for PEI can be attributed to its higher chelate effect and pKa.

The use of membranes among other technologies for the removal of arsenic from water is reviewed by Shih (2005) and Choong *et al.* (2007), and described in detail in Chapter 11.

10.6 FUTURE PERSPECTIVE

Due to the progress in membrane science and technology membrane separation processes have become popular treatment alternatives to many applications in water systems including metals removal. From the economical point of view these processes are still more costly than other traditional separation methods, especially for large-scale applications. However, when a tight and reliable removal of metals is required, membrane separation may be the preferred choice. There are a few limitations related to the use of membranes as indicated in Section 10.4. In addition, even tight membrane processes such as RO have not proved to be an absolute barrier against a variety of trace compounds (Snyder *et al.* 2007). Therefore, future treatment schemes may integrate a combination of processes such as sorption, chelation, and ion exchange, together with membrane separation processes, incorporating membrane chemical structure changes to increase their selective action. The technological formula that ensures absolute protection from trace compounds has not yet been defined, and therefore should be further investigated.

KEY POINTS

Membrane separation processes are considered emerging technologies capable of improving dramatically traditional processes such as granular filtration, precipitation, disinfection, biological treatment, and to assist in complete elimination of trace pollutants. Within the broad family of membrane separation processes there are "loose" membrane-based technologies, the separating mechanism of which is mainly size exclusion. These technologies include Microfiltration (MF) and Ultrafiltration (UF), which are operated successfully under relatively low levels of pressure. The other type of technologies that can be termed "tight" membrane separation processes require higher levels of pressure to enable free water passage with simultaneous rejection of particulate and even dissolved materials. This group includes Nanofiltration (NF) and Reverse Osmosis (RO), which are also termed desalination. For these two processes, the separation is based not only on physical mechanism related to membrane pore size (relatively small) that serve as a barrier, but also on the chemical structure of membrane material that can dissolve, attract, or reject various substances. The advance in membrane science and technology and their high and reliable separation efficiency, make this class of processes very suitable for the removal of heavy metals and related substances.

REFERENCES

Chilyumova E. and Thöming J. (2008). Nanofiltration of bivalent nickel cations – model parameter determination and process simulation. *Desalination*, **224**(1–3), 12–17.

Choong T. S. Y., Chuah T. G., Robiah Y., Gregory Koay F. L. and Azni I. (2007). Arsenic toxicity, health hazards and removal techniques from water: an overview. *Desalination*, **217**(1–3), 139–166.

de la Rubia A., Rodríguez M., León V. M. and Prats D. (2008). Removal of natural organic matter and THM formation potential by ultra- and nanofiltration of surface water. *Water Res.*, **42**(3), 714–722.

Favre-Réguillon A., Lebuzit G., Murat D., Foos J., Mansour C. and Draye M. (2008). Selective removal of dissolved uranium in drinking water by nanofiltration. *Water Res.*, **42**(4–5), 1160–1166.

Gubbuk I. H., Gungor O., Alpoguz H. K., Ersoz M. and Yılmaz M. (2010). Kinetic study of mercury (II) transport through a liquid membrane containing calix[4]arene nitrile derivatives as a carrier in chloroform. *Desalination*, **261**(1–2), 157–161.

Košutić K., Furač L., Sipos L. and Kunst B. (2005). Removal of arsenic and pesticides from drinking water by nanofiltration membranes. *Sep. Purif. Technol.*, **42**(2), 137–144.

Kozlowski C. A. and Walkowiak W. (2002). Removal of chromium(VI) from aqueous solutions by polymer inclusion membranes. *Water Res.*, **36**(19), 4870–4876.

Landaburu-Aguirre J., García V., Eva Pongrácz E. and Keiski R. (2006). Applicability of membrane technologies for the removal of heavy metals. *Desalination*, **200**(1–3), 272–273.

Lee S., Lee E., Ra J., Lee B., Kim S., Choi H. S., Kim S. D. and Cho J. (2008). Characterization of marine organic matters and heavy metals with respect to desalination with RO and NF membranes. *Desalination*, **221**(1–3), 244–252.

Li H., Gao Y., Pan L., Zhang Y., Chen Y. and Sun Z. (2008). Electrosorptive desalination by carbon nanotubes and nanofibres electrodes and ion-exchange membranes. *Water Res.*, **42**(20), 4923–4928.

Lytle D. A., Sarin P. and Snoeyink V. L. (2005). The effect of chloride and orthophosphate on the release of iron from a cast iron pipe section. *J. Water Supply: Res. Technol., AQUA*, **54**, 267–281.

Mbarecka C., Nguyenb Q. T., Alaouib Q. T. and Elaboration D. B. (2009). Characterization and application of polysulfone and polyacrylic acid blends as ultrafiltration membranes for removal of some heavy metals from water. *J. of Hazard. Mater.*, **171**, 93–101.

Mimoune S., Belazzougui R. E. and Amrani F. (2007). Purification of aqueous solutions of metal ions by ultrafiltration. *Desalination*, **217**(1–3), 251–259.

Nanda D., Tung K.-L., Hsiung C.-C., Chuang C.-J., Ruaan R.-C., Chiang Y.-C., Chen C.-S. and Wu T.-H. (2008). Effect of solution chemistry on water softening using charged nanofiltration membranes. *Desalination*, **234**(1–3), 344–353.

Sang Y., Li F., Gu Q., Liang C. and Chen J. (2008). Heavy metal-contaminated groundwater treatment by a novel nanofiber membrane. *Desalination*, **223**(1–3), 349–360.

Shih M.-C. (2005). An overview of arsenic removal by pressure-driven membrane processes. *Desalination*, **172**(1), 85–97.

Snyder S. A., Adham S., Redding A. M., Cannon F. S., DeCarolis J., Oppenheimer J., Wert E. C. and Yoon Y. (2007). Role of membranes and activated carbon in the removal of endocrine disruptors and pharmaceuticals. *Desalination*, **202**(1–3), 156–181.

Tang Z., Hong S., Xiao W. and Taylor J. (2006). Impacts of blending ground, surface, and saline waters on lead release in drinking water distribution systems. *Water Res.*, **40**, 943–950.

Van der Bruggen B., Koninckx A. and Vandecasteele C. (2004). Separation of monovalent and divalent ions from aqueous solution by electrodialysis and nanofiltration. *Water Res.*, **38**, 1347–1353.

Velizarov S., Matos C., Oehmen A., Serra S., Reis M. and Crespo J. (2008). Removal of inorganic charged micropollutants from drinking water supplies by hybrid ion exchange membrane processes. *Desalination*, **223**(1–3), 85–90.

Zamariotto D., Lakard B., Fievet P. and Fatin-Rouge N. (2010). Retention of Cu(II)– and Ni(II)–polyaminocarboxylate complexes by ultrafiltration assisted with polyamines. *Desalination*, **258**(1–3), 87–92.

Chapter 11

Arsenic removal processes

Zdravka Lazarova, Sabrina Sorlini, Frausta Prandini and D. Staniloae

11.1 INTRODUCTION

Contamination of drinking waters with toxic metals is much more widespread than previously believed, and levels that were once considered safe are now known to be health threats. In some areas, the natural high arsenic content of the drinking-water has caused chronic arsenic poisoning. The current needs to remove arsenic from drinking water is now a world problem, apart from the well publicised critical situation in Bangladesh and India, and so must be addressed as a matter of great urgency (Robins *et al.* 2001).

Arsenic, which is a widely distributed element in the earth's crust, occurs in drinking waters as inorganic compounds (arsenate and arsenite), and/or organic species. The different arsenic compounds have different levels of toxicity: the inorganic compounds, for example, are of a higher health concern than the organo-arsenicals which are typically only a small percentage of the arsenic pool in groundwater; the trivalent arsenic species are more mobile in the environment, and 25–60 times more toxic than the pentavalent (Korte & Fernando, 1991).

It is known that there is no a unique method for removal of toxic metals such as arsenic from drinking water because many factors influence its selective separation from the natural water matrix. The urgent need to find efficient removal techniques for arsenic in drinking waters has focused on physical and chemical processes. The choice of a suitable treatment method and its practical implementation depends first of all on the composition of the drinking water: concentration and oxidation state of the toxic metal, aqueous pH-value, presence and concentrations of other competitive ions. Arsenic has different oxidations states, and occurs in water as tri- and/or pentavalent compounds. Various species of the same valence can be formed depending on the aqueous pH-value: for example, arsenate (V) exists as a mono- or bivalent anion in the pH-range from 4 to 10, whereas arsenite (III) is a neutral species. Other important characteristics are TDS-concentration (total dissolved solids including minerals, salts, metals, cations and anions), presence of competing ions (sulphate, phosphate, fluoride, silicate, carbonate etc.), and their amount in the contaminated water.

11.2 AVAILABLE TECHNOLOGIES AND IMPLEMENTATION

There are a number of commercially available methods that are suitable for removal of toxic metals such as arsenic from drinking water: chemical co-precipitation, adsorption, ion exchange, and membrane filtration Jekel & Amy, 2006). From practical point of view, however, not only the efficiency of the treatment method but also the treatment costs should be taken into consideration. Furthermore, the appropriate treatment technology has to be easy in operation and maintenance, and user friendly. Thus, in spite of the existence of many removal processes, the development of efficient and cheap technologies for selective removal of toxic metals from drinking water, and for disposal of metal bearing wastes still represents a challenging task for the chemical and environmental engineers.

11.2.1 Oxidation

Reduced inorganic As(III) (arsenite) should be converted to As(V) (arsenate) to improve its removal with some processes like chemical precipitation and ion exchange. So, the arsenic oxidation should be accomplished by providing an oxidizing agent before the processes for arsenic removal. Chlorine, permanganate, ozone and granular manganese dioxide media are highly effective for this purpose, as they are capable to provide an arsenic oxidation higher than 95%. Chlorine dioxide and monochloramine are ineffective in oxidizing As (III). Ultraviolet (UV) light, by itself, is also ineffective, but if the water is spiked with sulphite, UV photo-oxidation can offer a good As(III) conversion (Ghurye & Clifford, 2001; USEPA, 2003).

11.2.2 Precipitation

The method is reliable, and therefore the most widely used for treatment of arsenic containing waters. The treatment procedure consists of several successively following process steps: coagulation by addition of chemical reagents (usually metal salts) and rapid mixing followed by flocculation, sedimentation, and filtration of the precipitate. Bi- or tri-valent salts of aluminium or iron, which are water-hydrolysable, have been found to be appropriate as coagulants in certain pH-ranges: ferric hydroxide – in the wide pH-range between pH 5 and pH 11, and aluminium coagulants – in the relatively narrow pH-range from pH 6 to pH 7.2 (Jekel & Amy, 2006).

In general, however, optimized coagulation-filtration systems are capable of achieving over 90% removal of As(V) and iron-based coagulants, including ferric sulphate and ferric chloride, are more effective at removing As(V) than their aluminum-based counterparts; this is because iron hydroxides are more stable than aluminum hydroxides in the pH range 5.5 to 8.5 (USEPA, 2003).

The main disadvantage of the coagulation/co-precipitation method is the use of chemicals (metals salts) which has to be added to the treated water to cause co-precipitation of the toxic metals. The resulting residue is rich in heavy metals and generates a bulky sludge which has to be treated further or safely disposed (typical values that can be used to estimate the quantity of sludge are: alum 0.33 kg dry sludge/kg alum; 0.48 kg dry sludge/kg ferric chloride; 0.59 kg dry sludge/kg ferric sulphate – (Crittenden *et al.* 2005). The process is very sensitive to the presence of many anions (silicates, phosphates, sulphates etc.) which can reduce significantly the removal efficiency. Moreover, the method is not appropriate for removal of neutral species such as As(III) (Jekel & Amy, 2006).

Another treatment method, which is very similar to the coagulation/co-precipitation with metal salts, is the **Lime Softening**. The only difference lies in the kind of the chemicals added to the treated water. Lime softening is a two step coagulation/co-precipitation procedure consisting of:

Step 1: Addition of *lime* $Ca(OH)_2$ to increase the pH-value (pH $\sim$ 10) and precipitate out the carbonate hardness in the water. The lime hydrolyses, and forms $CaCO_3$ which adsorbs *As* ions;

Step 2: Addition of Na_2CO_3 to eliminate the non-carbonate hardness.

The lime softening method is applicable for very hard waters (the optimal value is at pH $\sim$ 11). In spite of the large coagulant dose required (800–1200 mg/l), the *As*-concentration can not be reduced to the maximum concentration limit, therefore secondary treatment is required. The resulting large volume of sludge contains calcium arsenate which converts into $CaCO_3$ in the landfill disposal sites releasing *As* back into the environment (Meenakshi & Maheshwari, 2005).

Sustainable handling options for arsenic containing sludges have been described recently by Oberacker *et al.* (2003).

11.2.3 Adsorption

The removal of toxic metals from water by adsorption is a very important process. Recently, an excellent critical review on the arsenic removal from water/wastewater using adsorbents has been published based on more than 600 references (Mohan & Pittman Jr., 2007). The performance of various sorbents has been compared taking into consideration the specific sorption capacity towards arsenic compounds. The desorption of *As* from the loaded sorbent and its regeneration as well as the treatment/disposal of the arsenic concentrates are briefly discussed.

The adsorption is based on the forces of attraction between the metal ions and the surface charges of porous solids. Various materials with high surface area, synthetic or natural, are able to act as sorbents for *As*: activated carbon, activated alumina, fly ash, metal oxides (ferric, titanium, manganese, rare earth), sand coated with oxides, cellulose materials (sawdust, newspaper pulp), biological materials (living and non-living biomass; products of biological origin like chitin, chitosan).

Activated alumina and ferric oxide are the most widespread adsorbents for arsenic. The highly porous activated alumina has a large capacity towards arsenate in a very narrow pH-range (from 5.5–6.0). It needs relatively long residence time, and some ions (for example phosphate) disturb the adsorption process (Meenakshi & Maheshwari, 2005). The regeneration of a loaded activated alumina proceeds in two steps: flushing with 4% *NaOH* to replace *As* by *Na*, then flushing with acid to re-establish the H^+ on the alumina surface (Amy *et al.* 2000). The activated alumina loses 30–40% of its capacity during each regeneration step (Meenakshi & Maheshwari, 2005). The concentrated stripping solution has to be treated further.

Adsorbents containing iron offer a better choice for removal of metals like *As*. Iron shows strong affinity for both arsenic compounds, arsenate and arsenite, in a wide pH-range (from 5 to 10) (Meenakshi & Maheshwari, 2005). Various iron based adsorbents are known: Granular Ferric Hydroxide (β-*FeOOH*), ferric hydrite, geothite (α-*FeOOH*), zero valent iron. The granular iron hydroxide-based porous media *GFH* is able to binds strongly arsenic ions on its surface (5–6 time more efficient than the activated alumina), and the residual mass-produced is only 1/10th of that of the activated alumina (Pierce & Moore, 1992). *GFH* shows a long operating life because its high affinity for arsenate can't be affected by the presence of other common anions such as sulphate, bicarbonate, and chloride (Meng *et al.* 2000). Now, new commercial products of regenerable granular ferric hydroxide are available. For example, Streat *et al.* found a regeneration efficiency of 95–97% using 0.1 M NaOH (Streat *et al.* 2008).

Saha *et al.* (2001) have tested a lot of sorbents for removal of As(III) and As(V) from drinking water (some results are given in Table 11.1). It can be seen that the adsorption is a suitable method for removal of both arsenic species (pentavalent and trivalent) from water without any addition of chemicals. Generally, As(V) can be removed much easier and more efficiently than As(III). The best results have been achieved using GFH (hydrous granular ferric oxide), CalSiCo (iron coated coral limestone), and activated alumina.

Table 11.1 Removal of As from water with different sorbents (Saha *et al.* 2001).

Adsorbent		Removal %		Adsorbent		Removal %	
Composition	**Dose g/l**	**As (III)**	**As (V)**	**Composition**	**Dose g/l**	**As (III)**	**As (V)**
kimberlite tailing *(waste-diamond mining)*	10	25	40	sand	10	15	22
coal fly ash *(coal product)*	10	20	28	activated carbon	10	50	65
sawdust *(wood particles – saw in cutting)*	10	28	36	coated sand with iron-oxide	10	72	90
wood charcoal	10	19	37	Hematite (Fe_2O_3-mineral)	10	40	60
mushroom	10	22	35	granular ferric hydrous oxide	2	92	99
spent tea leaf	10	25	42	bauxite *(Al_2O_3-aluminium ore)*	10	58	80
banana pith	10	12	18	activated alumina	10	90	96
water hyacinth *(free-floating perennial plant)*	10	45	70	CaLSiCo *(coated coral limestone)*	5	90	98

11.2.4 Ion exchange

Ion exchange is an appropriate process for removal of arsenic from potable water. Anion exchangers loaded with *Cl*-ions are usually used, for example *Purolite* or *Amberlite* resins (Iesan *et al.* 2004). Vaaramaa and Lehto (2003) have tested different organic and inorganic ion exchangers and found unexpectedly that some cation exchangers are also able to remove arsenic from water.

During the removal process using anion exchangers, the arsenic anions get exchanged for chloride ions of the resin. The arsenic concentration in water can be reduced to <1 µg/L at pH between 6.5 and 9.0 (Meenakshi & Maheshawari, 2005). Moreover, the resin bed can be reused after regeneration with saturated sodium chloride solution.

The conventional ion exchange process has some disadvantages (Meenakshi & Maheshawari, 2005). The most important of them are the low selectivity for arsenate and the strong influence of many competitive ions (sulphate, TDS, selenium, fluoride, nitrate) what leads to short operational life. The uncharged As(III) can not be removed by IE. Large volume of arsenic-laden brine is generated during the regeneration which has to be also treated.

Recently a study was published on the application of magnetically impregnated resins (MIEX) in slurry reactors as an alternative to fixed-bed adsorbers (Bourke & Nguyen, 2007).

11.2.5 Membrane filtration

A comprehensive overview of arsenic removal by pressure-driven membrane processes has been published recently by Shih (2005). Microfiltration is suitable for treatment of suspensions (particles with MW $>$50.000 or particle size $>$0.05 µm), therefore it can be applied for separation of particulate forms of arsenic, for example, after coagulation with ferric chloride or sulphate, or flocculation with cationic polymers. Ultrafiltration is generally not appropriate for arsenic removal. There are some new studies on the application of UF in combination with electric repulsion, or using special prepared negatively charged UF-membranes (Amy *et al.* 1998; Brandhuver & Amy, 2001).

The high pressure processes RO and NF show high efficiency for arsenic removal from waters. Efficiencies up to 98% have been documented by RO membrane processes (Ning, 2002). *As* (III) has to be pre-oxidized to be removed by RO. The Nanofiltration could be also an interesting treatment alternative to RO because NF-membranes operate at significantly lower pressures than does reverse osmosis, and offer inherently high product water fluxes (Kim *et al.* 2006; Gholami *et al.* 2006; Kosutic *et al.* 2005). The high-pressure membrane filtrations (NF, RO) have also some limitations. Because of the small pore size of the membranes used, lower percentage of product water than in the case of UF and MF can be produced; they are more energy consuming because higher pressure is needed to push the water through the small membrane pores; the membranes are prone to fouling, and expensive.

All membrane filtration processes produce two streams: purified water (permeate) and a concentrated solution called retentate. The retentate contains all the toxic metals removed from the water and represents a disposal problem.

Point-of-use (POU) type membrane filters can be used to meet the arsenic MCL in rural areas or some arsenic contaminated areas.

11.2.6 Novel removal methods

Recently a number of new methods or materials for removal of arsenic from drinking waters have been developed. Newcombe *et al.* (2006) have described co-precipitation (ferric chloride) integrated with moving bed active filtration. A new procedure for impregnation of iron onto granular activated carbon consisting of two steps (treatment of virgin activated carbon with ferrous chloride, followed by oxidation of ferrous to ferric iron) has been proposed by Deng *et al.* (2005). In pilot scale, removal of arsenic by dissolved air flotation has been studied (Kordmostafapour *et al.* 2006).

New materials have been developed for adsorption of arsenic from water: ion exchange fibers (Greenleaf *et al.* 2006), iron coated sponge (Nguen *et al.* 2006), chelating polymer resins loaded with Zr(IV), Ce(IV), Nd(III) (Dambies, 2004), hydrotalcite (Gillman, 2006).

Recently nano-sorbents have been applied for removal of heavy metals from water. It is known that nanoparticles (NP) can be excellent sorbents and catalysts with fast kinetics because of their high surface area due to porosity and small size. However, the industrial application of NP as sorbents is hampered by their small size because NP tend to suspend and flow away with the water or clog very fast the filters used to keep them inside the adsorption column. In fluidized beds, the NP escape from the reactor along with the fluidizing gas. Some researchers have tried to solve this problem by developing special NP-aggregates for water-purification processes. Apblett *et al.* (2007) have prepared aggregates of NP metal oxide by thermal decomposition of metal-organic complexes. For this purpose, a *Zn*-loaded resin (H-form of Dowex 650°C) has been pyrolysed at high temperature (560°C). The resulting porous *ZnO* agglomerates have an average diameter of 170 μm and high surfaces area (30.7 m^2/g). The *ZnO*-spheres are mechanically robust and have a capacity for arsenate adsorption of 985 μg As/g adsorbant.

The conventional polymeric anion exchange resins are mechanically stable but have low selectivity for arsenate, and short operation life especially at high concentrations of competing anions such as sulphates. Many iron hydroxide porous media have high affinity for arsenate and negligible affinity for other common anions but are mechanically unstable (Nort, 2005). Sylverster *et al.* (2007) have combined the above-mentioned advantages of these materials preparing a hybrid inorganic/organic sorbent (*ArsenXnp*). In this sorbent, the hydrous iron oxide nanoparticles (3–5 nm in diameter) are highly distributed in the porous bead of strong base ion exchange resin (300–1200 μm). The arsenic interacts with the nanoscale hydrous iron oxide rather than the anion exchange groups of the polymeric substrate. The new sorbent is characterised by high affinity to arsenic, rapid sorption kinetics, long operational bed life, and relatively effective regeneration.

Vatutsina *et al.* (2007) have impregnated fibrous polymeric ion exchangers with NP of hydrated ferric oxide. The new fibrous composite sorbent shows high capacity towards both arsenic compounds, pentavalent and threevalent, without any corrections of pH and/or pre-oxidation of *As*(III).

Recent developments of hybrid adsorption/membrane processes offer new alternatives for arsenic treatment (Ng *et al.* 2004).

Johnson (2004) has published a study on micellar-enhanced ultrafiltration of water containing arsenic. The toxic *As* has been adsorbed on aggregates consisting of positively-charged micelles of cationic surfactant and poly(ethylene oxide) chains. The big complex molecules have been removed from water by UF membranes with relatively large pores.

A novel treatment method for removal of *As* from water based on ion-exchange membrane separation and precipitation by bacteria has been studied by Oehmen *et al.* (2006).

Point of use treatment

The points of use devices are attractive for removing contaminants that pose only an ingestion risk, for example arsenic. The primary advantage of using POU treatment in a small system is the potential for reduced capital and treatment costs, relative to centralized treatment systems, making it a more economically viable alternative for smaller systems. However, this treatment process as suitable method for removing arsenic is not accepted by all states.

Non treatment options

Non treatment options, such as blending, connecting to a neighbouring water source and drilling a new well, can be used also to mitigate problematic arsenic level in drinking water. These options can be defined as follow:

- *Blending* – combine multiple water sources to produce a water stream with an arsenic concentration below the MCL.
- *Connecting to a neighboring water source* – purchase water that is below the MCL from a nearby system if an interconnection exists.
- *Drilling a new well* – abandon the old well and locate and install a new source. Drilling a new well may not be the best option if the aquifer has consistently high levels of arsenic.

The new source installations may or may not be more expensive than treatment. Assessments must be made on a case by case basis.

The easy way to choose a technology/How to choose the optimal technology?

When a treatment technology is selected the main considerations that must be taking into account must include water quality attributes (including pH levels and initial concentrations of iron, As(III) and As(V) present in water), ease of implementation with current system, management of residues and cost.

The following information represent a simplified overview of the considerations to take in account when a treatment technology is selected.

11.3 CONSIDERATION ON WATER QUALITY

Although arsenic occurs in natural waters both inorganic and organic forms, the preponderance are the inorganic forms such as arsenite [As(III)] and arsenate [As(V)]. The valence and species of inorganic arsenic are dependent on the pH and oxidation-reduction conditions of the water. The arsenate ions As(V) are removed more effectively from source waters that arsenite ions As(III) by iron coagulants, by precipitation of natural iron and by adsorptive media. Moreover, the arsenite ion As(III) is not removed by anion exchange resins because of its uncharged nature. Then, oxidizing As(III) to As(V) will result in a higher arsenic removal for water with predominately As(III). As a conclusion, the arsenic removal capacity is improved when As(III) is converted to As(V) using a strong oxidant and increasing levels of iron concentration are present.

The iron concentration in source water is one of the main drivers in technology selection and it is often less expensive than other arsenic removal technologies. Therefore, many of the most effective arsenic removal processes available are iron based treatment technologies such as chemical coagulation/filtration with iron salts and adsorptive media with iron based products. The iron level in the water source is a primary consideration in the selection of an optimal treatment technology because of the unique role that iron plays in facilitating arsenic removal.

The way the Fe:As ratio could influence the chosen treatment technology is shortly presented below.

- *HIGH iron levels (>0.3 mg/L), high Fe:As ratio (>20:1).* Iron removal processes can be used to promote arsenic removal from drinking water via adoption and co-precipitation.
- *MODERATE iron levels (>0.3 mg/L), low Fe:As ratio (<20:1).* If the iron to arsenic ratio in the source water is less than 20:1, then the addition of iron salts should be considered in a modified treatment process such as coagulation/filtration.
- *LOW iron levels (<0.3 mg/L).* In this case, technologies such as adsorptive media, coagulation/filtration, and ion exchange are best suited.

This process selection guide is very basic and the removal capacities depicted are meant to be used as a general "rule of thumb", and these removal capacities will only be achieved under optimum operational conditions for adsorptive and As(V) processes.

pH value adjustment is sometimes required in order to increase the arsenic removal capacity of adsorptive media in percent range of As(III) versus As(V) removed. Generally, pH value less than 7.0 is optimal for iron-based media adsorption and a pH value of 5.5 is optimal for alumina-based media adsorption. The active pH range for arsenic removal with iron oxides via coagulation/filtration related processes is 5.5 to 8.0. The As(V) removal by ion exchange processes is not significantly impacted by changes in pH.

11.4 TREATMENT PROCESS AND RESIDUALS MANAGEMENT

Other technical consideration must be taking into account when we analyze alternative arsenic treatment technologies. Some unit processes are more economically viable than others under specific circumstances. An ideal option for some utilities is to optimize the arsenic removal using the existing processes. Arsenic treatment systems produce a residual for disposal, as other treatment processes. Because the arsenic discharge options may be limited the handling and disposal methods must be considered prior to selecting a technology. The As concentration in some liquid waste residuals is high and usually hazardous and theses need either to be treated on site or taken to a disposal facility. However, most solid wastes associated with spent media are not hazardous and can be disposed of in landfills. State regulations vary from state to state for the discharge of residuals to water bodies and onto land. It is important to be aware of not only national regulatory requirements regarding residual management, but also applicable European requirements in order to better evaluate of existing practices and to plan for needed changes in treatment plant operations.

11.5 EXAMPLES OF REAL SCALE TREATMENT PLANTS FOR THE ARSENIC REMOVAL IN EUROPE

Italy

Arsenic is a significant contaminant of groundwater in many areas in Italy especially in central-north regions (Lombardia, Emilia Romagna, Veneto, Piemonte, Toscana & Lazio) where concentrations up to 300 µg/L are reached in groundwater.

Moreover, in 2003 in Italy, the 2001/31 Legislative Decree, that was the accomplishment of 98/83/UE DWD, also reduced the limit for arsenic in drinking water from 50 µg/L to 10 µg/L. This fact induced in drinking water facilities of these areas some technical impacts:

- many drinking water treatment plants managers needed to upgrade the existing plants where arsenic was previously already removed, in order to comply with the new lower limit;
- many drinking water treatment plants managers needed to build up new plants for arsenic removal when this contaminant was not previously a critical parameter.

Due to the importance of this problem, a working group called "Water for human consumption: arsenic removal in drinking water treatment plants" was activated in 2005. One activity of this working group was to perform an investigation about the main technologies adopted in Italy for arsenic removal in drinking water treatment plants. During this activity 18 plants were investigated.

Different kinds of technologies are used in Italy. In Table 11.2 the main characteristics of some plants for arsenic removal are shown. The most common solution is precipitation with iron and aluminium salts, but this technology doesn't assure to reach 10 µg/L when arsenic concentration in raw water is high. So, in these cases in the plants there is a tertiary treatment with *GFH* after the chemical precipitation. There are some plants that apply only adsorption technologies (*GFH* and TiO_2), but generally these solution is applied for small applications and with low initial arsenic concentration. Two plants are provided with ion exchange and two with reverse osmosis (RO). The RO technology, unlike the other analyzed solutions, is applied also for very high arsenic concentration in raw water, up to 100 µg/L; one of these plants is applied at point of use (POU).

Table 11.2 Drinking water plants for arsenic removal in Italy.

Technology	No of applications	Flow rate (L/s)	As concentration in raw water (µg/L)
Chemical precipitation	7	9.4–450	17–50
Chemical precipitation + GFH	2	5.5–26	44–65
GFH	6	2.6–30	15–65
TiO_2	2	0.58–11.1	30–50
Ion exchange	2	2–33	30
Reverse osmosis	2	0.33–34	50–100

As concerns the residues management, five of the seven plants adopting chemical precipitation have specific treatments for sludge: polyelectrolyte addition, sedimentation and dewatering. The other two plants discharge directly in sewerage system according to the limits of the Italian Legislative Decree 2006/152.

In the plants with adsorption and reverse osmosis the liquid residues are discharge in the sewerage system. In the case of ion exchange the brine is treated with $FeCl_3$.

Germany

At present, in Germany, 16 drinking water treatment plants apply the GFH-technique for treatment of *As* containing waters (Meenakshi & Maheshwari, 2005).

Denmark

Baastrup *et al.* (2008) presented a document where the aim was to determine if individual exposure to low levels of arsenic in drinking-water in Denmark is associated with a risk for cancer.

The study was based on the prospective Danish cohort Diet, Cancer and Health, which has been described in detail elsewhere (Tjønneland *et al.* 2007). In brief, 160,725 persons 50–64 years of age and living in one of 23 municipalities in the Copenhagen or Aarhus area were invited to participate. Of these, 57,053 persons (27,178 men and 29,875 women) accepted the invitation and were enrolled between 1993 and 1997. At enrollment, information was collected including on diet, beverages, smoking, education, medical conditions, occupations, reproductive factors, body mass index, and skin reaction to sun. The study "Diet Cancer and Health" has been approved by the relevant Scientific Committees and the Danish Data Protection Agency. Informed consent was obtained from all participants to search information from medical registers including the Danish Cancer Registry. Of the 57,053 cohort members, that are included 56,378 persons, who filled in the lifestyle questionnaire, reported daily intake of tap water, and had not had a cancer diagnosis before the enrollment.

Arsenic concentrations in Danish drinking water were obtained from a database managed by the Geological Survey of Denmark and Greenland (Thomsen *et al.* 2004). Information on the size and spatial location of 94 water supply areas was collected from local authorities and water utilities in 24 municipalities, covering the vast majority of the geocoded cohort addresses. The results show that there are some place in Aarhus and Odense municipality with arsenic concentration above 5 μg/L.

They found no statistical significant association between arsenic concentrations in Danish drinking-water and the risk for cancers of the lung, bladder, kidney, liver, prostate, or colorectum. The results indicated inverse associations between arsenic concentrations in Danish drinking-water and risk for skin cancers, suggesting that arsenic might have a protective effect at low concentrations. The results also indicated that arsenic in drinking-water might increase the risk for breast cancer.

Sweden

The Swedish Geological Institute of Sweden (SGU) reports what regions are at risk of high arsenic levels in drilled wells. This study shows that arsenic concentration is below 51 μg/L. In particular, five wells are arsenic concentration from 25 to 51 μg/L and eight wells with arsenic concentration from 10 to 25 μg/L. This means that some wells could be a problem for arsenic contamination for drinking purpose.

Netherlands

Het Waterlaboratorium (Haarlem, Netherlands) collected data, in 2006, on arsenic concentration in 71 wells. It showed that arsenic levels in groundwater are below 10 μg/L and only two places have a concentration of 13 μg/L in Breda–Dorst and 17.8 μg/L in Loosdrecht.

Latvia

There are no data about arsenic concentration in water in Latvia. Only a study of Gosk *et al.* (2007) shows that arsenic concentrations exceed the quality standards (of 10 μg/L) in 1% of the samples.

Romania

There is a small number of data about As in drinking water in Romania. Concentrations out of limits are finding out in mineral water sources (e.g. Sarul Dornei/Furnica source –0.5 up to 2.05 mg/L) and surface water (e.g. Valea Nogajului river – up to 0.054 mg/L), especially in the old volcanoes region. There are not treatment plants with special processes for As removal.

KEY POINTS

- The main source of arsenic in groundwater is due to the natural release of this metalloid
- Arsenic is a toxic metalloid
- Inorganic constituents of arsenic (arsenate and arsenite) are of higher health impact than organic compounds.
- Arsenic has different oxidation states and it occurs in water as tri- and/or pentavalent compounds.
- Specific standard value for arsenic in drinking water applied by the WHO, EPA and UE is 10 μg/L.
- Physical and chemical processes are applied for arsenic removal in drinking waters: chemical co-precipitation, adsorption, ion exchange, and membrane filtration.
- Reduced inorganic As(III) should be converted to As(V) to improve its removal.
- High Fe:As ratio promotes arsenic removal.
- Arsenic can be removed at centralized, small and household scale.
- In order to define the best technology, water quality, semplicity of implementation, management of residues and cost should be considered.

REFERENCES

Amy G. L., Edwards M., Benjamin M., Carlson K., Chwirka J., Brandhuber P., McNeil L. and Vagliasindi F. (1998). Arsenic Treatability Options and Evaluation of Residuals Management Issues, Draft Report, April 1998, AWWARF.

Amy G., Edwards M., Brandhuber P., McNeill L., Benjamin M., Vagliasindi F., Carlson K. and Chwirka J. (2000). Arsenic Treatability Options and Evaluation of Residuals Management Issues. AWWARF and AWWA, Denver, Colorado.

Apblett A. W., Kurijavar S. I., Bagbas A. and Piquette A. P. (2007). Nanotechnology in drinking water supply. In: Session Sustainability in Water Supply: Advances in Adsorption Processes for Drinking Water Treatment, B. J. Martinas, Q. Li and T. H. Nguyen (eds), Proceedings the 233rd ACS National Meeting, Chicago, IL, March 25–29, pp. 1567.

Brandhuber P. and Amy G. (2001). Arsenic removal by a charged ultrafiltration membrane – Influences of membrane operating conditions and water quality on arsenic rejection. *Desalination*, **140**, 1–14.

Bourke M. F. and Nguyen H. V. Inorganic contaminant removal from water, US Patent 7291272 (11/06/2007).

Crittenden J. C., Trussel R. R., Hand D. W., Howe K. J. and Tchobanoglous G. (2005). Water Treatment: Principles and Design. John Wiley & Sons, Hoboken, New Jersey.

Dambies L. (2004). Existing and prospective sorption technologies for the removal of arsenic in water. *Sep. Sci. Technol.*, **39**, 603–627.

Deng B., Caviness M. and Gu Z. (2005). Arsenic removal by activated carbon-based materials. In: Advances in Arsenic Research, P. A. O'Day, D. Vlassipoulos, X. Meng and L. G. Bening (eds), Chapter 20, ACS Symposium Series, *Am. Chem Soc.*, Vol. 915, pp. 284–293, Washington, DC.

Gholami M. M., Mokhtari M. A., Aameri A. and Alizadeh Fard M. R. (2006). Application of reverse osmosis technology for arsenic removal from drinking water. *Desalination*, **200**, 725–727.

Gillman G. P. (2006). A simple technology for arsenic removal from drinking water using hydrotalcite. *Sci. Total Environ.*, **366**, 926–931.

Greenleaf J. E., Lin J. C. and Sengupta A. K. (2006). Two novel applications of ion exchange fibers: Arsenic removal and chemical-free softening of hard water. *Environ. Prog.*, **25**, 300–311.

Iesan C., Bapat S. S., Fries B., Coman D. and Florea D. (2004). Arsenic removal from drinking water by ion exchange resins. *Environm. Eng. Manage. J.*, **3**, 283–291.

Jekel M. and Amy G. L. (2006). Arsenic removal during drinking water treatment. In: Interface Science in Drinking Water Treatment, G. Newcombe and D. Dixon (eds), Elsevier Ltd., Amsterdam, Chapter 11, pp. 193–206.

Johnson M. 2004. Removal of Arsenic and Other Heavy Metals from Water using Micellar- Enhanced Ultrafiltration. Thirteenth Annual Undergraduate Research International Symposium, Arizona, NASA Space Grant Consortium, Tucson, April 24.

Kang M., Kawasaki M., Tamada S., Kamei T. and Magara Y. (2000). Effect of pH on the removal of arsenic and antimony using reverse osmosis membrane. *Desalination*, **131**, 293–298.

Kim D. H., Kim K. W. and Cho J. (2006). Removal and transport mechanisms of arsenics in UF and NF membrane processes. *J. Water Health*, IWA Publ., 215–223.

Kosutic K., Furac L., Sipis L. and Kunst B. (2005). Removal of arsenic and pesticides from drinking water by nanofiltration membranes. *Sep. Purif. Technol.*, **42**, 137–144.

Kordmostafapour F., Pourmoghadasm H., Shahmansouri M. R. and Parvaresh A. (2006). Arsenic removal by dissolved air flotation. *J. Appl. Sci.*, **6**, 1153–1158.

Korte E. and Fernando Q. (1991). A review of As(III) in ground water. *Q. Crit. Rev. Environ. Control*, **21**, 1–39.

Meenakshi P. and Maheshwari R. C. (2005). Arsenic removal from water. *Asian J. Water, Environ. Pollut.*, **3**, 133–139.

Meng X. G., Bang S. B. and Korfiatis G. P. (2000). Effect of silicate, sulphate and carbonate on arsenic removal by ferric hydroxide. *Water Res.*, **34**, 1255–1261.

Mohan D. and Pittman Jr. C. U. (2007). Arsenic removal from water using adsorbents – critical review. *J. Hazard. Mater.*, **142**, 1–53.

Newcombe R. L., Hart B. K. and Moeller G. (2006). Arsenic removal from water by moving bed active filtration. *J. Environ. Eng.*, **132**(1), 5–12.

Ng K. S., Ujang Z. and Le-Clech P. (2004). Arsenic removal technologies for drinking water treatment. *Rev. Environ. Sci. Biotechnol.*, **3**, 43–53.

Nguyen T. V., Vigneswaran S., Ngo H. H., Pokhrel D. and Viraraghaven T. (2006). Specific treatment technologies for removing arsenic from water. *Eng. Life Sci.*, **6**, 86.

Ning R. Y. (2002). Arsenic removal by reverse osmosis. *Desalination*, **143**, 237–241.

Nort K. P. (2005). Attrition loss analysis for arsenic adsorption media. Sandia National Laboratory Report SAND2006-0374.

Oberacker G. F., Maier D. and Maier M. (2003). In: Vom Wasser, M. Jekel (ed.), Vol. 100, Wiley-VCH Verlag, Weinheim, Germany.

Oehmen A., Viegas R., Velizarov S., Reis M. A. M. and Crespo J. G. (2006). Removal of heavy metals from drinking water supply through the ion exchange membrane bioreactor. *Desalination*, **199**, 405–407.

Pierce L. M. and Moore C. B. (1982). Adsorption of arsenite and arsenate on amorphous iron hydroxide. *Water Res.*, **16**, 1247–1253.

Robins R. G., Nishimura T. and Singh P. (2001). Removal of arsenic from drinking water by precipitation, adsorption or cementation. In: M. F. Ahmed, M. A. Ali and Z. Adeel (eds), Proc. BUET-UNU Intern. Workshop on Technologies for Arsenic Removal from Drinking Water, BUET, Dhaka, Bangladesh, May 5–7, 59–68.

Saha J. C., Dikshit K. and Bandyopadhyay M. (2001). Proc. BUET-UNU International Workshop on Technologies for Arsenic Removal from Drinking Water, Dhaka, Bangladesh, May 5–7, 76–84.

Shih M. C. (2005). An overview of arsenic removal by pressure driven membrane processes. *Desalination*, **172**, 85–97.

Streat M., Hellgardt K. and Newton N. L. R. (2008b). Hydrous ferric oxide as an adsorbent in water treatment. Part 2. Adsorption studies. *Process Saf. Environ. Protect.*, **86**, 11–20.

Sylvester P., Westerhoff P., Möller T., Badruzzaman M. and Boyd O. (2007). A hybrid sorbent utilizing nanoparticles of hydrous iron oxide for arsenic removal from drinking water. *Environ. Eng. Sci.*, **24**, 104.

USEPA. Arsenic Treatment Technology Evaluation Handbook for Small Systems, EPA 816/R-03/014, July, 2003.

Vaaramaa K. and Lehto J. (2003). Removal of metals and anions from drinking water by ion exchange. *Desalination*, **155**, 157–170.

Vatutsina O. M., Soldatov V. S., Sokolova V. I., Johann J., Bissen M. and Weissenbacher A. (2007). A new hybrid (polymer/inorganic) fibrous sorbent for arsenic removal from drinking water. *React. Funct. Polym.*, **67**, 184–201.

Chapter 12

Hybrid processes

Zdravka Lazarova

12.1 DESCRIPTION OF TECHNOLOGY

To meet the increased demands for drinking quality-water, effective and low cost treatment technologies are needed. One way is to improve the existing technologies for removal of metals and related substances from water. Another possibility is to combine existing methods into the so-called hybrid process. If several different physical, chemical and/or biological processes are tied together so that they complement each other well, this is called a hybrid process.

Hybrid membrane treatment

The most known hybrid application is the combination of pressure driven membrane separations with high capacity sorbents. When used as a single process, MF and UF membranes can only remove particles and bacteria but do not reject dissolved metal ions and most of the organic (colour, NOM, synthetic organic chemicals). There is a possibility to incorporate adsorbents, coagulants or activated sludge into the treated aqueous solution (feed) to increase the removal performance of the membrane separation.

Another approach involves bonding the dissolved metals to a special bonding agent. Such an innovative process is the *Polyelectrolyte-Enhanced Ultrafiltration* (PEUF), based on the complexation of the negatively charged arsenate ions with the positively-charged polyelectrolyte (Sabatini *et al.* 2004). Humic substances and chitosan also make stable complexes with the heavy metals which can be removed from water due to the complex retention (Verbych *et al.* 2005).

MEUF (*Micellar-Enhanced Ultrafiltration*) is another effective low-pressure technique for metal removal based on incorporation of the metal ions within surfactant-made micelles (Iqbal *et al.* 2007). Cationic surfactants form micelles with positively charged surface when its concentration is higher than the critical micelle concentration. Arsenic, which exits in water as either arsenite or arsenate, can be adsorbed on the surface of the micelles by electrostatic interaction, and the metal loaded micelle can be separated from the aqueous stream by a membrane with an appropriate pore size.

The use of sorbents containing multiple polymeric functional groups is a novel technique to achieve high metal sorption capacity under convective flow conditions. Such sorbents are formed by attachment of various polyamino acids (MW: 2500–10,000), such as polyaspartic acid (cation sorption), polyarginine (oxyanion sorption), and polycysteine (chelation exchange), directly on the membrane pore surfaces (Ritchie & Bhattacharyya, 2002). Since these sorbents have high selectivity over non-toxic metals, such as calcium, they are ideal candidates for hybrid processing with RO/NF.

The recent development of *Submerged Hybrid Membrane Systems* offers alternative hybrid technologies for arsenic treatment (Ng *et al.* 2004). The membrane in hybrid systems allows better phase separation between particles binding arsenic and the treated water than those obtained in traditional separation systems like settlement. The membrane bioreactor (MBR) process is standing as the best example of the success of hybrid membrane treatment. The current tendency is to apply *Immersed Membrane Filtration* where the membrane is directly submerged in the reactor.

Hybrid biological treatment

An innovative alternative technology represents the removal of both trivalent and pentavalent arsenic species from groundwater by biological iron oxidation. (Katsoyaiannis *et al.* 2004). Arsenic is removed by direct adsorption or co-precipitation on the pre-formed biogenic iron oxides. Iron oxidation is catalysed by several microorganisms, which are indigenous in most groundwater (such as *Gallionella ferruginea* and *Leptothrix ochracea*). A biofilm is formed during the aeration of the fixed-bed adsorber (with polystyrene beads as filtration media). The bacteria accumulate in the filtration media and grow up by obtaining their energy either by the oxidation of ferrous iron (*Galionella*) or by consumption of organic matter (*Leptothrix*). The main product of the biological oxidation of iron is usually a mixture of iron oxides containing significant amounts of organic matter. The intermixing of iron oxides, organic material and bacteria produces multiple sorbing solids with unique metal retentation properties.

Other hybrid methods

Coagulation/Precipitation of arsenic with $FeCl_3$ is combined with a moving bed active filter (Newcombe *et al.* 2006). The waste effluent, using 10% of influent for transport, is retained in a clarifier for settling prior to the water recycling. Research observations support the hypothesis that the formation and renewal of iron oxide-coated sand in the active filter is a viable mechanism for high efficiency arsenic removal.

A new nano-level magnetic filtration/adsorption process for purifying water containing high concentrations of arsenic is developed using a supported surface complex of natural magnetite $FeOFe_2O_3$ (Shaban *et al.* 2003). In the presence of an external magnetic field, a synergistic effect was observed in using supported magnetite. It is due to the high gradient magnetic separation effect, as colloidal particles with satisfactory magnetic properties are present in natural water systems.

12.2 IMPLEMENTATION OF TECHNOLOGY FOR THE REMOVAL OF HEAVY METALS AND RELATED SUBSTANCES

Combined adsorption-ultrafiltration process for removal of Cr(III) ions from water is proposed by Pagana *et al.* (2008) in which porous ceramic membranes and γ-Al_2O_3 nanoparticles as sorbents are used. Both adsorption and UF, take place simultaneously in the internal surface of the membrane tube. The membrane can be regenerated after heating followed by washing.

The removal efficiency of microfiltration (MF), which is much lower than that of nanofiltration (NF) due to its larger pore size, can be increased by adding a small amount of nanoscale zero valent iron (nZVI). When an amount of 0.1 g/L of nZVI is added into the treated arsenic solution, the removal of As(V) by MF increases from 40% up to 90%, and that of As(III) from 37% to 84% (Nguyen *et al.* 2009).

Lazarova (2006) performed MEUF of zinc containing waters using a water-soluble polyvinyl electrolyte (PIES) which forms complexes with the zinc cations. Poly-electrolyte concentration of about 1.5 g/L is sufficient to enable the full Zn-rejection (initial Zn-concentration of 100 mg/L).

Removal of arsenate from groundwater by MEUF using cationic surfactants CPC (hexadecylpyridinium chloride) and CTAB (hexadecyltrimethyl-ammonium bromide) is developed by Iqbal *et al.* (2007). The UF with a regenerated cellulose membrane is followed by post-treatment of the effluent with PAC (powdered activated carbon), filtration or settling before discharging to the environment.

Surfactant-enhanced hybrid PAC/CFMC (Powdered Activated Carbon/Crossflow MF) of Ni is developed by Aydiner *et al.* (2006). The surfactant incorporates either on the surface or within the pores of the membrane and leads to better rejection of Ni. The flux declines as much as 10 times, whereas Ni rejection increases two times in comparison to the non-surfactant-enhanced hybrid process.

An improved degree of removal of manganese ions from groundwater by UF is achieved upon addition of the chelating polymer PAA (polyacrylic acid) into the feed water (Choo *et al.* 2007). Sulphate ions present in the feed water reduce the manganese rejection efficiency due to the Donnan exclusion, while hardness competes with manganese in the chelation reactions with PAA.

The environmental friendly biopolymer chitosan can be applied for simultaneous complexation and enhanced ultrafiltration of heavy metals (Verbych *et al.* 2005). Chitosan-metal complexes allow metal ions to be rejected without any substantial reduction of membrane productivity. The performance of coupled chitosan/UF for arsenic removal is studied by Lin *et al.* (2008).

The complexation of Pb(II), Cu(II), Ni(II) and Co(II) with HS (humic substances) leads to their effective removal by cellulose acetate UF-membrane (Alpatova *et al.* 2004). This is an effective method for simultaneous removal of HS and heavy metals harnessing the principle of reagent-reinforced ultrafiltration. The degree of heavy metal rejection improves with an increase in HS–Me ratio due to increase in completeness of metal binding. The metal rejection increases with pH increase, and in neutral/alkaline solutions it exceeds 80%.

A novel treatment method for removal of arsenic and mercury is the Ion Exchange Membrane Bioreactor (IEMB) process (Oehmen *et al.* 2006). The method is a combination of Donnan dialysis (with charged ion exchange membranes, which excludes similarly charged ions, and permit the permeation of oppositely charged ions), and biological process. Anion exchange membrane is used for the removal of *As*, and a cation exchange membrane for *Hg*. In the bio-compartment (separated by the membrane), the arsenic anions are precipitated by sulphate-reducing bacteria as As_2S_3, whereas Hg^{2+} are converted into Hg^0 by suitable bacteria. The IEMB process has a high potential for use in drinking water treatment systems, and offers numerous advantages over currently implemented processes minimizing the risk of secondary pollution.

The use of biological oxidation of Fe and Mn proved to be a promising technique for supplementary arsenic removal from contaminated groundwater (Katsoyaiannis *et al.* 2004). The method is based on the growth of certain species of indigenous bacteria which are capable of oxidizing the soluble iron and manganese ions; the oxidized forms can be subsequently removed from the aqueous streams by over 97%, through their transformation to insoluble oxides and separation by a suitable filter medium.

A novel approach to remove arsenic and iron from water is the so called constructed soil filter (CSF) in which As(III) is oxidized to As(V) by filter media via natural oxidation, and arsenic is co-precipitated with iron (Nemade *et al.* 2008). Water containing arsenic, iron, and phosphate is passed through the CSF media.

The oxidation takes place due to presence of various oxides such as iron, manganese, aluminium, and the microbial diversity in the media of soil bioreactor.

12.3 ADVANTAGES AND DISADVANTAGES

The innovative integration of physical, chemical, and biological processes into one "hybrid" process has many advantages. In comparison to the conventional "single" processes for removal of metals and related substances from ground and surface water, the application of hybrid processes leads to higher separation selectivity and efficiency, less energy consumption, and smaller footprint. More economical and environmental friendly water treatment technologies can be applied.

The advantages of the hybrid processes containing membrane separation include lower operating costs and better fouling control over the conventional membrane systems.

The combination of the physicochemical adsorptive treatment with biological processes avoids the use of chemicals for oxidation; does not require monitoring of the breakthrough point as in common sorption processes because the sorbents (for example iron oxides) can be continuously produced in situ; there is no need of regeneration and replacement of adsorbing material.

12.4 CASE STUDIES

Pilot-scale experiments are performed by Ujang *et al.* (2004) on the removal of high amount of arsenic and other contaminants from drinking water using Immersed Membrane Filtration (IMF) and its combination with Powdered Activated Carbon (PAC) or zeolite (ZEO). Hybrid membrane systems with PAC demonstrated better performances on arsenic removal. IMF coupled with PAC (5 mg/l) removed 41% of total arsenic, compared to 20% for zeolite (5 mg/l). With the introduction of adsorbent, IMF can be also applied for rejection of organics dissolved in water.

Arsenic is removed from water in pilot scale by coagulation with polyaluminium chloride and Dissolved Air Flotation (Kordmostafapour *et al.* 2006). For the separation stage, a new hybrid process of flotation and membrane separation is developed by integrating specially designed submerged microfiltration modules directly into the flotation reactor (Blöcher *et al.* 2003). In this way the advantages of both flotation and membrane separation are used to overcome their limitations. The feasibility of this hybrid process is proven using powdered synthetic zeolites as bonding agents. Stable fluxes of up to 80 $l\ m^{-2}\ h^{-1}$ were achieved with ceramic flat-sheet multi-channel membranes at low transmembrane pressure (<100 mbar). The concentration of all toxic metals containing in the treated water, namely copper, nickel and zinc, is reduced from initial concentrations of 474, 3.3 and 167 mg l^{-1}, respectively, to below 0.05 mg l^{-1}, consistently meeting the discharge limits.

Dual Porosity Filtration (DPF) is a novel technology designed to remove suspended fines (colloidal – 100 μm size) and dissolved contaminants from water, using sedimentation, adsorption and bio-degradation as unit operating processes (Cederkvist *et al.* 2010; Jensen *et al.* 2011). The DPF facility consists of a filter of layers with high-flow compartments sandwiched horizontally between low-flow compartments, containing a filter bed consisting of calcite grains (diam. 1–3 mm). The filter is approximately 10 m wide, 50 m long and 10 cm high. The DPF-facility has shown to be effective in retention of cationic metals like Cu, Pb and Zn. To gain arsenate (AsO_4^{3-}) and chromate (CrO_4^{2-}) reactivity of the filter bed, an in-situ coating procedure based on iron oxides and humic substances was performed. Iron oxides, in the form of ochreous sludge were added to a recycling volume of water in the DPF-filter, followed in the next step by addition of humic substances. To dissolve the ochreous sludge and later re-precipitate the iron as oxides at the calcite surface under co-precipitation with humic substances, a microbiologically driven reducing-oxidizing process was carried out, using sucrose as carbon-source. By the addition of chromate- and arsenate-pulses before, during and after the coating procedure, the retention capacity of the filter was assessed. Results showed that the coating definitely improve the reactivity of the filter towards the two oxyanions with outlet-concentrations well below 10 ppb. Interestingly, a strong affinity is achieved already upon the embedding of ochreous sludge. This suggests that DPF combined with ochreous sludge and calcite holds a strong chromate and arsenate treatment potential.

Three new technologies for arsenic removal, advanced ion-exchange operations with indefinite brine recycling (AIXO-IBR), microsand-assisted oxidation-adsorption (MAOA), and coagulation-assisted ceramic filtration (CCF), are tested in Scottsdale and Tucson (Arizona), and Billings (Montana) (Galeziewski *et al.* 2002). Preliminary design parameters and cost curves are developed based on results of a year-long pilot-scale testing. In general, the capital cost of these three technologies ranges from $500,000 to $1,200,000 for a one million gal/day (mgd) facility, excluding building and land costs, with annual operation and maintenance (O&M) costs being $75,000 to $200,000. Capital and O&M costs are proportional to system complexity, with the MAOA and CCF processes being more economical than the AIXO-IBR system.

Pushpavanam *et al.* (2005) proposed an electrochemical method for removal of arsenate from drinking water, wherein the arsenate is removed by adsorption of metal hydroxide, formed by 'in-situ' anodic oxidation. The method obviates the drawbacks of the commonly used physicochemical treatment processes. The electrochemical device consists of an electrochemical cell fitted with an anode of mild steel or aluminum plate and stainless steel cathode with an inter-electrode distance of 0.5 to 1.5 cm. Drinking water containing 0.5 to 3.0 mg/L of arsenate at pH in the range from 3 to 10 and temperature between 20–60°C is electrolyzed at anode and cathode current densities between 0.05–0.2 $A \cdot dm^{-2}$. The iron hydroxide/aluminum hydroxide formed from the anode during electrolysis adsorbs the arsenate present in the water and settles at the bottom. The removal efficiency of this method is up to 98%.

A pilot-scale Membrane Polyelectrolyte-Enhanced Ultrafiltration System with spiral wound membrane is applied which can treat 4 gpm of water with 100 ppb of arsenic in a continuous mode and removes more than 90% of the arsenic (Sabatini *et al.* 2004). It is based on the complexation of the negatively charged arsenate ions with the positively-charged polyelectrolyte QUAT (poly diallyldimethyl-ammonium chloride). The complex is removed from water using an ultrafiltration membrane with a MWCO of 10,000 smaller than that of the polyelectrolyte. The produced water (permeate) is free of polyelectrolyte and contains a greatly reduced concentration of arsenic. The retentate, containing arsenate and QUAT in high concentration, is sent to a crystallizer where a multivalent cation (e.g. barium) is added to precipitate the arsenate. The arsenate salt is settled out and the polymer-rich supernatant recycled, producing a very low volume solid waste. For sulphate concentrations lower than 10 ppm, the treatment cost per gal is less than other technologies (e.g. adsorption and ion exchange technologies).

12.5 FUTURE PERSPECTIVES

Energy, raw materials, and water – everything is scarce and therefore expensive. This market offers plenty of room for new ideas. Treatment processes aimed at purification of the water resources and producing drinking water with high quality are in demand so as never before. Improving the established or developing novel low-cost water treatment technologies based on the advantageous hybrid processes is an emerging issue.

KEY POINTS

The integration of different physical, chemical and/or biological processes into the so-called "hybrid" process creates new advantageous applications in the removal of metals and related substances from ground and surface water. The most known hybrid application is the combination of pressure driven membrane processes such as MF and UF with high capacity sorbents or special metal-bonding agents (polyelectrolytes, surfactants, polyamino acids, natural biopolymers such as chitosan) which substantially increases the separation selectivity and efficiency. Pilot-scale tests have been conducted and promising results achieved of arsenic removal with Immersed Membrane Filtration and its combination with powdered activated carbon as well as with Membrane Polyelectrolyte-Enhanced Ultrafiltration.

The membrane bioreactor process is the best example of the success of hybrid membrane treatment. The hybrid system allows better phase separation between particles binding arsenic and the treated water than those obtained in traditional separation systems.

REFERENCES

Alpatova A., Verbych S., Bryk M. and Nigmatullin R. (2004). Ultrafiltration of water containing natural organic matter: heavy metal removing in the hybrid complexation-ultrafiltration process. *Sep. Purif. Technol.*, **40**(2), 155–162.

Aydiner C., Bayramoglu M., Kara S., Keskinler B. and Ince O. (2006). Nickel removal from waters using surfactant-enhanced hybrid PAC/MF process. I. The influence of system-component variables. *Ind. Eng. Chem. Res.*, **45**(11), 3926–3933.

Blöcher C., Dorda J., Mavrov V., Chmiel H., Lazaridis N. K. and Matis K. A. (2003). Hybrid flotation – membrane filtration process for the removal of heavy metal ions from wastewater. *Water Res.*, **37**(16), 4018–4026.

Cederkvist K., Holm P. E. and Jensen M. B. (2010). Full-scale removal of arsenate and chromate from water using a limestone and ochreous sludge mixture as a low-cost sorbent material. *Water Environ. Res.*, **82**(5), 401–408.

Choo K. H., Han S. C., Choi S. J., Jung J. H., Chang D., Ahn J. H. and Benjamin M. M. (2007). Use of chelating polymers to enhance manganese removal in ultrafiltration for drinking water treatment. *J. Ind. Eng. Chem.*, **13**(2), 163–169.

Galeziewski T. M., Kwan P., Kommineni S., Dotson A. and Johnson B. (2002). Three new arsenic removal technologies: how to design them and how much they will cost. Annual AWWA Conference and Exposition Proceedings, June 16–20, New Orleans, Louisiana, 957–973.

Iqbal J., Kim H. J., Yang J. S., Baek K. and Yang J. W. (2007). Removal of arsenic from groundwater by micellar-enhanced ultrafiltration (MEUF). *Chemosphere*, **66**, 970–976.

Jensen M. B., Cederkvist K., Bjergager P. E. R. and Holm P. E. (2011). Dual porosity filtration for treatment of stormwater runoff: first proof of concept from Copenhagen pilot plant. *Water Sci. Technol.*, **64**, 1547–1557.

Katsoyaiannis I. A. and Zouboulis A. I. (2004). Application of biological processes for the removal of arsenic from groundwaters. *Water Res.*, **38**, 17–26.

Katsoyaiannis I. A., Zouboulis A. I., Althoff H. and Bartel H. (2002). As(III) removal form groundwaters using fixed-bed upflow bioreactors. *Chemosphere*, **47**, 325–332.

Kordmostafapour F., Pourmoghadas H., Shahmansouri M. R. and Parvaresh A. (2006). Arsenic removal by dissolved air flotation. *J. Appl. Sci.*, **6**(5), 1153–1158.

Lazarova Z. (2007). Removal of metals from drinking water – overview. Proc. EU-COST 637 Program, Antalya, Turkey, *Oktober, 2007.*

Lin C. F., Wu C. H. and Lai H. T. (2008). Dissolved organic matter and arsenic removal with coupled chitosan/UF operation. *Sep. Purif. Technol.*, **60**, 292–298.

Newcombe R. L., Hart B. K. and Moller G. (2006). Arsenic removal from water by moving bed active filtration. *J. Environ. Eng.*, **132**(1), 5–12.

Nemade P., Kadam A. M. and Shankar H. S. (2008). Arsenic and iron removal from water using constructed soil filter – a novel approach. *Asia Pac. J. Chem. Eng.*, **3**(5), 497–502.

Ng K. S., Ujang Z. and Clech P. L. (2004). Arsenic removal technologies for drinking water treatment. *Rev. Environ. Sci. Bio/Technol.*, **3**, 43–53.

Nguyen V. T., Vigneswaran S., Ngo H. H., Shon H. K. and Kandasamy J. (2009). Arsenic removal by a membrane hybrid filtration system. *Desalination*, **236**(1–3), 363–369.

Oehmen A., Viegas R., Velizarov S., Reis M. A. M. and Crespo J. G. (2006). Removal of heavy metals from drinking water supplies through the ion exchange membrane bioreactor. *Desalination*, **199**(1–3), 405–407.

Pagana A. E., Sklari S. D., Kikkinides E. S. and Zaspalis V. T. (2008). Microporous ceramic membrane technology for the removal of arsenic and chromium ions from contaminated water. *Microporous Mesoporous Mater.*, **110**, 150–156.

Pushpavanam S., Sozhan G., Mohan S. and Vasudevan S. (2005). *U.S. Pat. Appl. Publ.*, Council of Scientific and Industrial Research, India.

Ritchie S. M. C. and Bhattacharyya D. (2002). Membrane-based hybrid processes for high water recovery and selective inorganic pollutant separation. *J. Hazard. Mater.*, **92**(1), 21–32.

Shaban I. S., Creager S. E., Karanfil T. and Navratil J. D. (2003). Magnetic filtration/adsorption for removal of arsenic from drinking water. Abstracts of Papers, 225th ACS National Meeting, New Orleans, LA, United States, March 23–27.

Sabatini D. A., Gallo H. D., Acosta E. and Scamehorn J. F. (2004). Removal of arsenic from drinking water using polyelectrolyte-enhanced ultrafiltration (PEUF), Abstracts of Papers, 228th ACS National Meeting, Philadelphia, PA, United States, August 22–26, 2004.

Ujang Z., Ng K. S. and Henze M. (eds) (2004). Comparative study on immersed microfiltration (MF) and hybrid systems with powdered activated carbon (PAC) and zeolite (ZEO) on removal of high concentration of arsenic. Environmental Biotechnology: Advancement in Water and Wastewater Applications in the Tropics, Selected Proceedings of the IWA International Conference on Environmental Biotechnology, Kuala Lumpur, Malaysia, Dec. 9–10, 2003, Meeting Date 2003, 323–330.

Verbych S., Bryk M., Alpatova A. and Chornokur G. (2005). Ground water treatment by enhanced ultrafiltration. *Desalination*, **179**(1–3), 237–244.

Lightning Source UK Ltd.
Milton Keynes UK
UKOW010146281112

202812UK00004B/125/P